PHYSICAL

GEOGRAPHY
LAB MANUAL   THIRD EDITION

J. Steven Kite
Amy E. Hessl
West Virginia University

Views from the Mountain State

Kendall Hunt
publishing company

Cover image © Shutterstock.com

Kendall Hunt
p u b l i s h i n g   c o m p a n y

www.kendallhunt.com
Send all inquiries to:
4050 Westmark Drive
Dubuque, IA 52004-1840

Contents

Lab Exercise 1

Maps, Globes, and Time Zones

Reading assignment that should be completed **BEFORE** Lab 1

Readings

Strahler, A.H., 2013, **Introducing Physical Geography**, John Wiley & Sons, New York, **Introduction** & **Chapter 1,** *or alternative reading assigned in GEOG 107.*

Materials Used in Lab

- 12 inch globe showing political boundaries
- World Atlas maps
- Large Mercator projection world maps
- Glossary in Strahler or alternative physical geography textbook

Learning Objectives

Upon successful completion of this lab, the student should be able to:

- Conceptualize distances expressed in metric units
- Convert distance measurements from metric to imperial units and vice-versa
- Find a location on a map or globe using latitude, longitude coordinates
- Define the latitude, longitude coordinates for a location
- Become familiar with the lab facility, course syllabus, class structure, instructor, and fellow Physical Geography lab students

Math Tutorial assignment that should be completed **BEFORE** Lab 1

Math Tutorial

GEOG 106 Math requirements should not be a hurdle for most students because they are covered in high school Algebra I (9th grade standards in most states).

However, math skills become rusty with disuse, so some may benefit from refresher tutorials available at this great website written just for introductory geoscience students:

Wenner, Jennifer, and Baer, Eric, *2019*, **The Math You Need, When You Need It—*Math tutorials for students in introductory geosciences*.** https://serc.carleton.edu/mathyouneed/index.html (*viewed May 2019*).

The Math You Need, When You Need It website has eight available tutorial topics, but the following four are most relevant to GEOG 106. The topics are listed in the order in which they are encountered during the semester:

- How do I change units on a number? *Unit conversions in the geosciences.* https://serc.carleton.edu/mathyouneed/units/index.html
- How can I use topographic maps? *An overview of topographic maps and associated topics.* https://serc.carleton.edu/mathyouneed/slope/index.html
- How do I plot points on a graph? *Plotting geologic data in x-y space.* https://serc.carleton.edu/mathyouneed/graphing/plotting.html
- How do I calculate rates? *Calculating changes through time in the geosciences.* https://serc.carleton.edu/mathyouneed/rates/index.html

Quiz 1 will include questions to assess your mastery of these basic math skills, and all four skills will be used to complete Exercises 1 and 2, so check out these web resources ASAP, ideally before coming to class to begin Lab Exercise 1.

Metric and Imperial Measurement Systems

The ***metric system*** is the preferred system of measurement for all nations of the world except Myanmar, Liberia and the United States. Also called the ***International System of Units***, or "SI units", the metric system's use in the U.S. was permitted by the Metric Act of 1866. Many still have not embraced it over 150 years later, but most geoscientists and geographers exclusively use the metric system. SI units are easy to use since they are based on a single fundamental unit for each physical quantity (length, area, volume, mass, etc.). Smaller and larger units are related to the fundamental unit by factors or powers of 10, so one can convert by simply moving the decimal place.

The ***U.S. Customary System of Units***, adapted from the British Imperial System, has dozens of weird, hard-to-remember conversion factors. This point can be demonstrated by asking a group of friends "How many inches are in the 120 yard total length of a football field?" If it takes your friends longer than five seconds to answer, then it took longer than it would take people from almost any other country to calculate that there are 10,000 centimeters in a 100 meter soccer pitch.

Exact metric-imperial unit conversions can be vital to engineers and scientists, but it's easier for most of us to ***"think metric"*** than to convert. What's a meter? One meter is about 39.3701 inches, but true understanding is more than numbers. Most readers grew to 1 meter tall between the ages of 3 and 4½ years. Adult U.S. females average about 1.62 meters tall, while U.S. male adults average about 1.76 meters. Three meters is just short of the height of a basketball hoop. Less than one person in 10 million can sprint 100 meters in under 10 seconds. The walk from Brooks Hall to the Mountainlair is about 180 meters, while a one-way ride on the whole WVU PRT route is about 5,800 meters (= 5.8 kilometers). It sounds cool to drive 110 kilometers/hour in Canada, but it is 1 mile/hour below the speed limit on I-79. On a different scale, a drive from WVU to the U.S. Capital is 342,000 meters (342 kilometers).

Common metric units for linear measurements:

millimeter (mm)	0.001 meters	"milli" prefix means 1×10^{-3}
centimeter (cm)	0.01 meters	"centi" prefix means 1×10^{-2}
meter (m)	1.0 meter	no prefix means $1 = 1 \times 10^{0}$
kilometer (km)	1000.0 meters	"kilo" prefix means 1×10^{3}

Common metric–imperial conversion factors (approximate, unless otherwise noted):

1 cm = 0.3937 inches	1 inch = 2.540 cm (exactly)
1 m = 3.281 feet	1 foot = 0.3048 m
1 km = 0.621 miles	1 mile = 1.609 km

Numerous on-line apps help with unit conversions, but to stress the role of computation in geoscience we will make a few unit conversions from imperial to metric and metric to imperial using a calculator.

1. Convert the following units using the conversion factors from the previous section. Round results to three *significant digits*. *(If your understanding of significant digits is rusty, please ask your instructor for a refresher.)*
   a. 9.25 miles = _____ km
   b. 38.1 cm = _____ inches
   c. 63 inches = _____ cm
   d. 34,000 mm = _____ feet
   e. 4,862 feet = _____ m

Latitude and Longitude

Latitude is the angular distance north or south of the equator measured from the center of the Earth (see figure 1-1). Lines designating these angles run east to west, parallel to the equator on maps and globes. A **parallel** is a line that connects all points along the same latitude. A **great circle** is any circle that can be drawn around the Earth's circumference whose center coincides with the center of the Earth. A **small circle** is any circle whose center does not coincide with the Earth's center. The **equator**, or equatorial parallel, has a physical basis as the only line of latitude that is also a great circle. All other parallels are small circles.

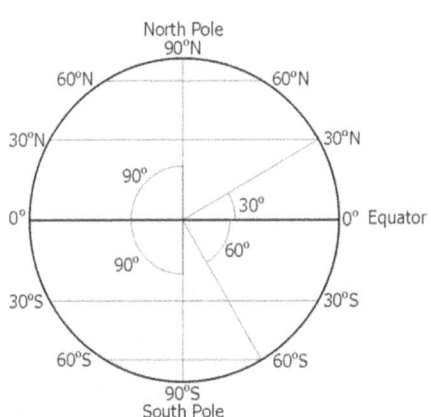

Figure 1-1
© *Kendall-Hunt Publishing Company*

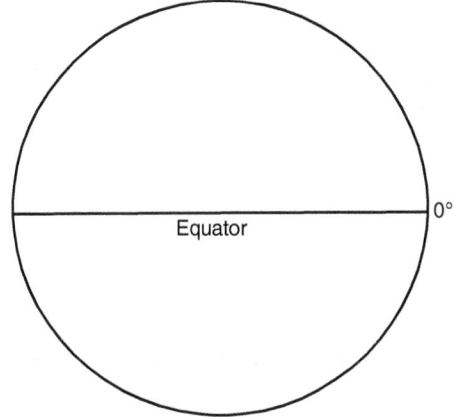

Figure 1-2
© *Kendall-Hunt Publishing Company*

Figure 1-1 depicts a cross-section of the Earth if it were bisected (cut in half) along the polar axis. The angle of latitude is measured from the center of the Earth to the surface, north or south of the equator, and extends from 0° at the equator to 90° at the poles.

2. Use figure 1-2 and a protractor to draw the 40° north parallel. Be sure to label the parallel. Use your atlas to name a city that lies along this parallel.

Longitude is the angle east or west of an arbitrarily selected point on the Earth's surface (see figure 1-3). *Meridians* connect all points along the same longitude. All meridians are lines along great circles that run north to south from the North Pole to the South Pole. Meridians converge at the poles. Other locations were used as the "zero" meridian in the past, but since 1884 the internationally recognized *Prime Meridian* has passed through The Royal Observatory in Greenwich, England, a suburb of London.

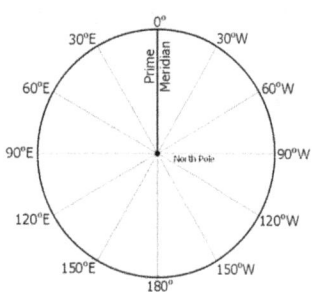

Figure 1-3
© *Kendall Hunt Publishing Company*

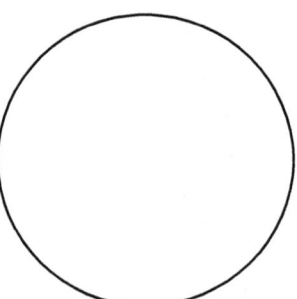

Figure 1-4
© *Kendall Hunt Publishing Company*

Figure 1-3 depicts a cross-section of the Earth as if it were bisected at the equator. The angle of longitude is measured from the center of the Earth, east or west of the prime meridian, and extends from 0° at the Prime Meridian to, a 180° meridian exactly halfway around the world from the Royal Observatory.

3. Use figure 1-4 and a protractor to draw the prime meridian and the 180° longitude meridian. Draw the 80° west meridian. Be sure to label the meridians and the hemispheres.

4. Consult a globe to name a city that lies along the 80° west meridian.

5. Consult a globe to identify an important time zone boundary that coincides with 180° longitude throughout much of the Pacific Ocean.

6. Both cities have a 40° N longitude, so you might incorrectly think the shortest route from Morgantown, WV, USA, to Beijing, China, would be due west or due east. Using the globe and a very taut piece of string, lay out the great circle route between Morgantown, WV, USA, and Beijing, China. What geographic features will this route pass over?

Maps, Globes, and the Geographic Grid

7. Use the globe to locate these cities. Provide their geographic coordinates (i.e. latitude and longitude) to within 1 degree. Do not forget to note the hemisphere by putting N, S, E, or W after each coordinate.

City	Latitude	Longitude
Morgantown, WV, USA	40° N	80° W
Denver, CO, USA		
London, England, UK		
Beijing, China		
Cape Town, South Africa		
Baghdad, Iraq		

8. If you move to a place that is 20½° farther south and 75° west of Morgantown, what would be your new coordinates and where would you live?

9. What is an *antipode*? Give the antipode for Morgantown, WV.

10. Find the north and south *magnetic poles*. Give their latitude and longitude, as shown on the globe.

11. Where is the southernmost land in the world? (Hint: It is ice covered.) Give its latitude, and explain why one can't assign it a unique longitude.

12. Find the northernmost point of land in the world. Give its latitude and longitude.

13. Use the globes and maps provided in the atlas to complete the table. Look in atlas tables to find the actual surface areas of the three countries.

Country	Size rank (1, 2, or 3) based on . . .		Actual Area (km²)
	Mercator Map	Globe	
Greenland			
India			
Saudi Arabia			

14. Why do you think differences in size exist between the globe's representation and the Mercator projection's representation of the three countries?

Time Zones (see your world atlas or textbook)

Long ago, communities set their clock as they saw fit: often so the sun was at its highest point in the sky at 12:00 noon. Real problems arose with the late 19th Century advent of rapid train transportation and nearly instantaneous telegraph communication. Imagine a train conductor trying to keep on schedule when every station was on a different clock! To reduce the chaos, the United States adopted four "standard" time zones in 1883.

15. If a WVU sporting event starts at 7:05 p.m. in Morgantown, West Virginia, at what local time would it stream live in the following cities? *(Ignore the complication of daylight savings time, which is not used in Hawaii.)*

Place	Time difference from Greenwich time (a.k.a. Coordinated Universal Time)	Difference from Morgantown	Local Time for Start of Broadcast
London, England	0	+5:00	12:05 am
Paris, France			
New York City, NY			
Chicago, IL			
Denver, CO			
Los Angeles, CA			
Anchorage, AK			
Honolulu, HI			

16. How many time zones exist in the United States today, excluding U.S. Territories, such as Puerto Rico and Guam? How many in the contiguous United States?

17. What is the International Dateline?

18. Why is the International Dateline important when considering global time zones?

19. Assume that you are an international businessperson who needs to get to Japan for an important meeting. If you fly west from San Francisco, what time and date issues should you consider when booking your flight?

20. Why does the International Dateline zigzag around easternmost Siberia, the westernmost islands of the Aleutians, and (since 1997) the Pacific islands in the nation of Kiribati? (See your world atlas).

21. What location celebrates the New Year (12:00 midnight on December 31st/January 1st) earlier than any other place on Earth?

Topographic Maps and Aerial Photos

Reading assignments to be completed <u>BEFORE</u> Lab 2

Readings

Strahler and Strahler, **<u>Introducing Physical Geography</u>**, John Wiley and Sons, New York.
Read Appendix 3, Topographic Map Symbols.
Wenner and Baer, 2019, How can I use topographic maps? Viewed May 2019 at https://serc.carleton.edu/mathyouneed/slope/index.html.
Wenner and Baer, 2019, How do I construct a topographic profile? Viewed May 2019 at https://serc.carleton.edu/mathyouneed/slope/topoprofile.html.

Materials Used in Lab

- U.S. Geological Survey Topographic Map Symbols sheet, available from https://pubs.usgs.gov/gip/TopographicMapSymbols/topomapsymbols.pdf
- Morgantown North (W.VA.–PA.) and Devils Tower (WY) 1:24,000 scale USGS topographic maps
- 10 squares/inch graph paper
- Map wheels and Rulers marked in mm and tenth of inches
- Hand Calculators
- *Parts of the exercise can be completed using the USGS TopoView website: https://ngmdb.usgs.gov/topoview/viewer/#4/39.98/-100.06*

Learning Objectives

Upon successful completion of this lab, students will be able to:

- Calculate distance using scale
- Understand how elevation is expressed using contour lines
- Calculate elevation change using contour lines
- Read and interpret physical features on aerial photos

Map Scale

A map is a representation of part of the Earth, commonly printed on a flat sheet of paper. *Map Scale* relates a measurement on a map to its corresponding real world distance. The many ways to show map scale include a *ratio scale* (e.g. 1:100,000) or a *representational fraction* (e.g. 1/100,000). Other common types of scales are the *bar scale*, and a *verbal scale* expressed as a simple equation or in words, such as "1 inch = 2000 feet" or "one centimeter equals 100 meters."

Maps are considered either large or small scale based on their level of detail and areal extent. *Large scale* maps depict smaller geographic areas, but are more detailed. *Small scale* maps depict larger geographic areas, but in less detailed. To help keep small scale straight from large scale, remember when map scales are written as fractions, such as 1/1,000,000 or 1/100, smaller scales yield smaller decimal fractions than larger scales: 0.000001 and 0.01 in this example.

1. Complete the following table by calculating distances on the Earth's surface represented by 1 inch or 1 cm on maps with four common U.S. map scales.

Map Scale	Real World distance shown by 1 inch on the map		Real World distance shown by 1 cm on the map	
1:1,000,000	ft	mi	m	km
1:250,000	ft	mi	m	km
1:62,500	ft	mi	m	km
1:24,000	ft	mi	m	km

2. Which of the four map scales best exemplifies a large scale map?

3. Which is a small scale map?

Topographic Maps

Topographic maps, or *"Topo Maps"*, illustrate the *topography*, or "the lay of the land" by means of *topographic contours.* Topographic contours are lines of equal elevation above sea level. Numerical elevations appear only on *index contours*, typically every fifth contour, which have bolder line weights than the more numerous **intermediate contours**. Buildings, urban areas, forests, and other features appear on these maps as special colors and symbols. The U.S. Geological Survey (USGS) produces most topographic maps used in this country, including most of the maps we will use. These four-sided USGS maps commonly are called *quadrangles,* or *"quads"*

Interpreting the Morgantown North Quadrangle

4. What is the map's scale, stated as a representational fraction?

5. Explain this map's scale in words.

6. One inch on the map represents _____ inches on the Earth surface, _____ feet on the Earth surface, or _____ miles on the Earth surface.

7. Give longitude and latitude of the southwest corner of the map.

   Longitude: _____ Latitude: _____

8. Give longitude & latitude of the northeast corner of the map.

   Longitude: _____ Latitude: _____

9. Based on your answers to questions 7 and 8, why are USGS maps at this scale, commonly called 7½ minute quadrangles?

10. Name the map that covers the area immediately to the south: _____. Name the map to the southwest: _____.

11. What is the *magnetic declination in degrees*, the difference between magnetic & true north, in the area? Why is declination essential in using a compass in the field?

Earth's magnetic poles are moving greatly. See current NOAA sources for details: e.g. https://www.ngdc.noaa. gov/geomag/calculators/magcalc.shtml#declination

12. What is the map's *contour interval* (elevation difference between successive contour lines)? _____

13. What is the *principal contour* interval (elevation difference between successive bold index contour lines)? _____

14. Using your familiarity with the WVU Downtown Campus, determine the approximate elevation of the ground in front of the Mountainlair student union?

15. How far is it from the Mountainlair to the WVU football stadium? Give your answer in feet, miles, meters, and kilometers.

16. What is the straight-line distance in meters from the source of West Run where it originates near the Pleasant Hill Cemetery (~1 km south-southeast of Exit 7 on I-68) to where it ends at its confluence with the Monongahela River?

What is the curvilinear distance in meters along the creek's circuitous path?

17. What is the *vertical relief* in feet (difference in elevation) between the Coliseum and the Monongahela River next to the WVU Core Arboretum?

Coliseum Elevation _____ft River Elevation _____ft Vertical relief _____ft

18. What general statement can be made about the spacing between contour lines in areas of varying topography? Where is the flattest (natural) land found on this map?

Topographic Profiles

A *topographic profile* is a drawing that shows the change in elevation of the land surface along a given line. It represents a cross-section of the real-world topography.

19. On graph paper provided in lab, draw a topographic profile from point A across Devils Tower to point B using a clip of the Devils Tower, WY, Quad, in figure 2-1.

- Use a *vertical exaggeration* (ratio of vertical scale to horizontal scale) of 10. You may either construct the topographic profile based on figure 2-1 or use the topographic map provided in class.

- Review the "How do I construct a topographic profile?" reading assignment if you are unsure how to draw the topographic profile.

- To make vertical exaggeration = 10, plot elevation using a vertical scale of 1 inch = 200 ft (=1:2,400 scale) and plot the horizontal distance using the same 1 inch = 2000 ft (1:24,000) scale as the map.

- The cross-section should be along a straight line from point A to Point B.

- Put point A on the LEFT side of the profile and point B is on the RIGHT side.

One way to plot the topo profile is to fold the graph paper along the top edge, and place the grid along the profile line on the topo map. At each contour-line crossing, mark a short dash on the top edge of the graph paper. Also mark streams and hilltops, where slope direction changes. Label the elevation of each mark as you go. Plot elevations as points on the graph paper. Connect points by a smooth best-fit curve and label major features. Draw a horizontal scale based on the width of each section = 5280 feet.

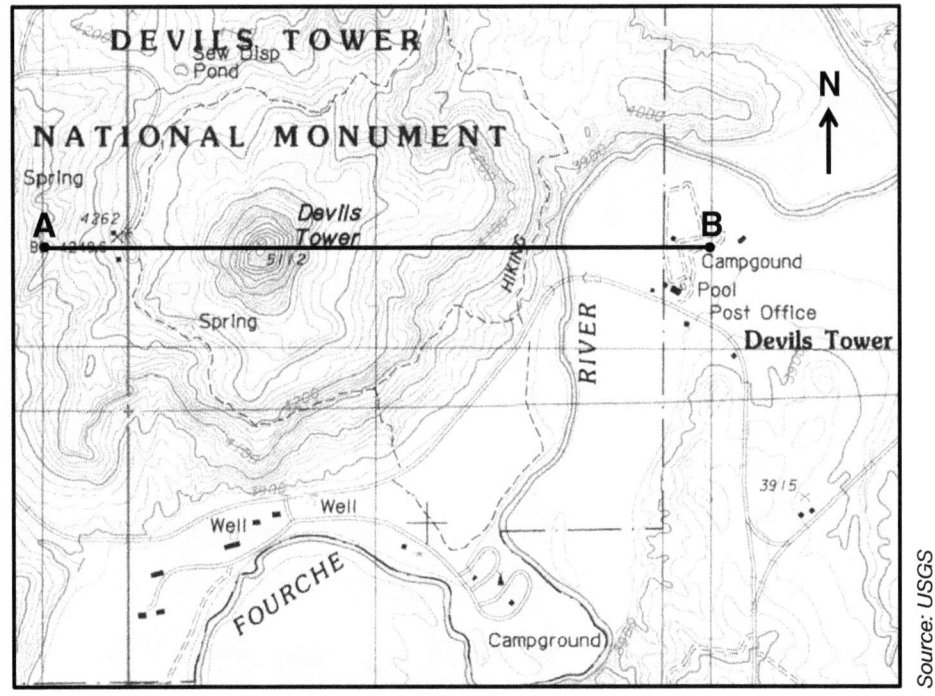

Figure 2-1 *Excerpt from Devils Tower, WY, Quadrangle.*

Topographic Maps from the Western U.S.

A checkerboard land-use pattern is obvious when flying over the American Great Plains. Topographic maps covering the western 3/4th of the country reveal a grid pattern inherited from the *United States Public Land System* (USPLS). *Sections* are 1 square mile grid squares used in the USPLS to survey property for most areas outside the original 13 states. Figure 2-2 shows how 36 sections are laid out in a regular pattern to make a *township*. Each township has an east-west coordinate called "*range*" and a north-south coordinate, confusingly also called "*township*".

Figure 2-2 *The United States Public Land System (USPLS) section pattern within a township has a standard number pattern, with section 1 in the northeast corner and section 36 in the southeast corner. The full pattern looks like this:*

6	5	4	3	2	1
7	8	9	10	11	12
18	17	16	15	14	13
19	20	21	22	23	24
30	29	28	27	26	25
31	32	33	34	35	36

Devils Tower Images

20. The Devils Tower 1:100,000 scale map in figure 2-3 shows Section 6, near the top of the map, lies due north of Devils Tower. Refer to figure 2-2 to determine which section the Devils Tower landform is located?

21. Figure 2-3 gives Devils Tower's elevation as 1558, but figure 2-1 shows it as 5112! Both maps are correct. Explain the discrepancy in tower elevation values.

22. For the next four images, which map and which air photo show more detail?

23. Which scale map and which scale air photo cover a larger area?

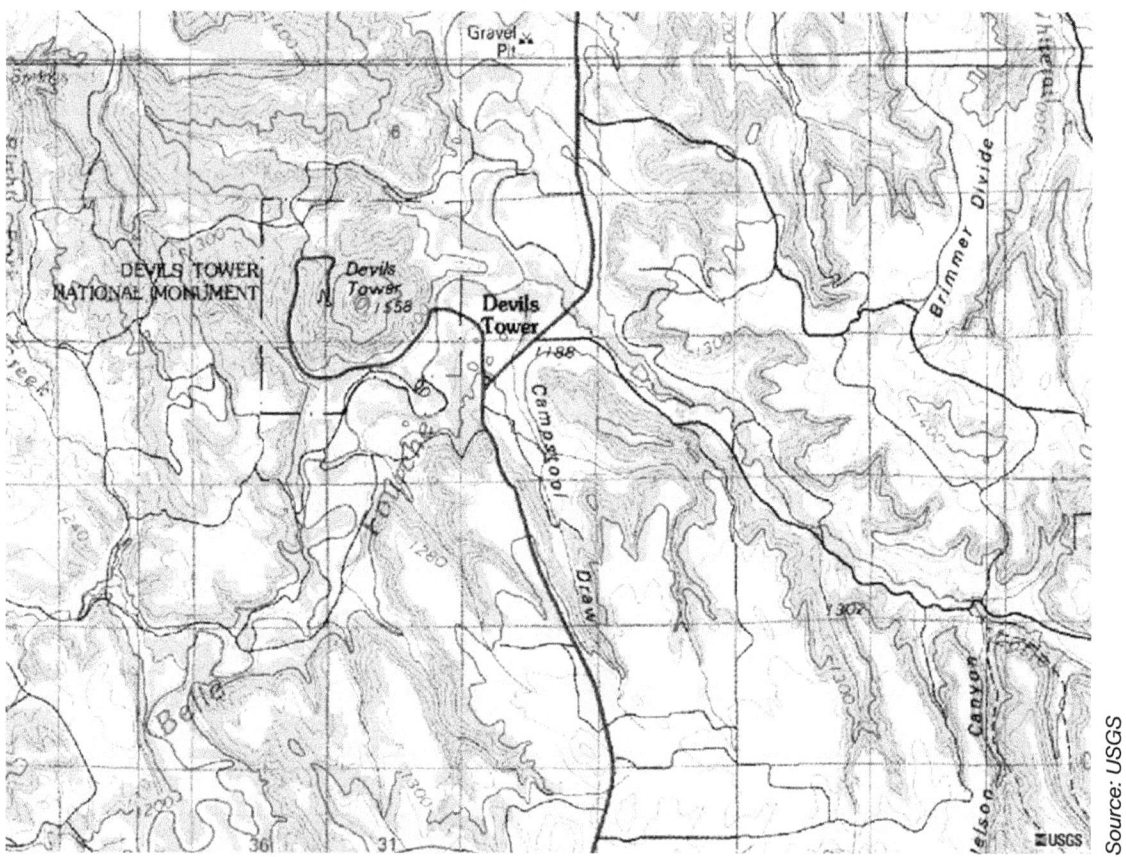

Figure 2-3   *Devils Tower topo map derived from USGS 1:100,000 scale map.*

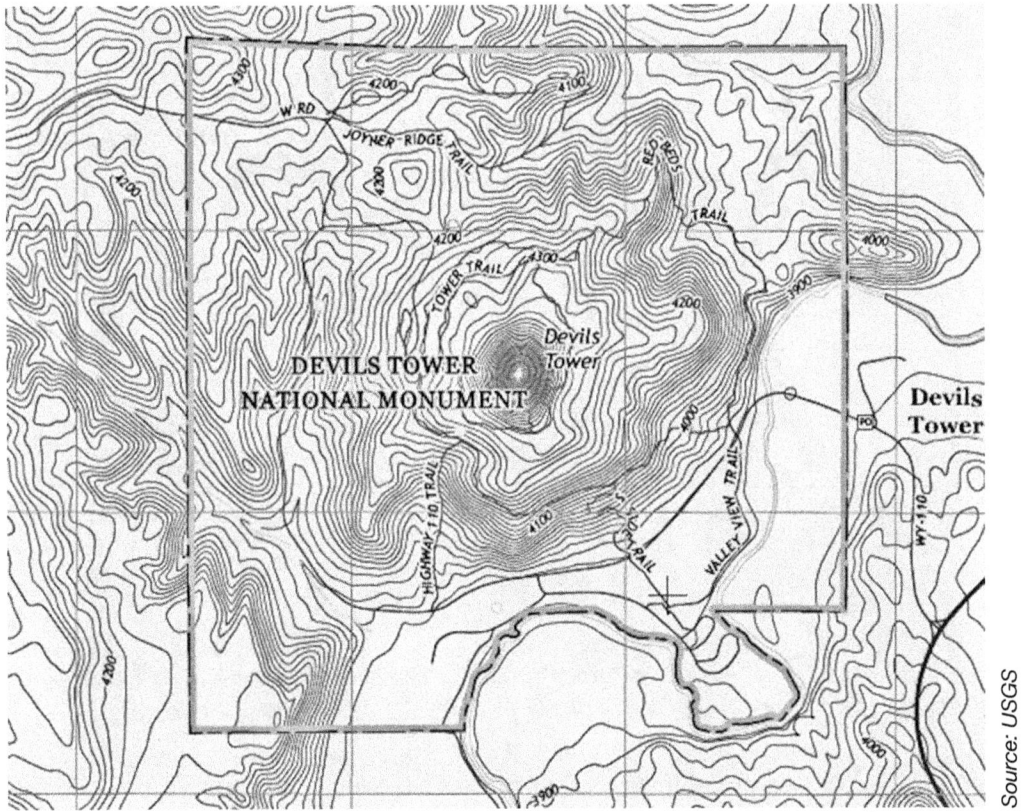

Source: USGS

Figure 2-4   *Devils Tower topo map derived from USGS 1:24,000 scale quad.*

Source: USGS

Figure 2-5   *Devils Tower aerial photo from USGS 1:100,000 scale image.*

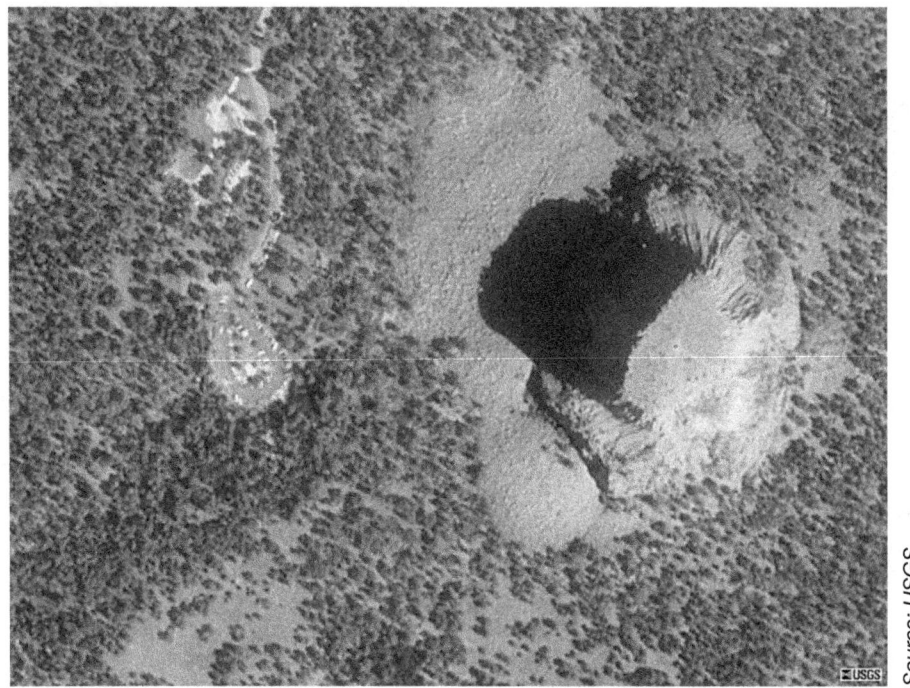

Figure 2-6   *Devils Tower aerial photo from USGS ~1:12,000 scale image.*

Source: USGS

24. North is toward the top of the image; use the tower's shadow direction as a sundial to estimate the time of day when the aerial photo in Figure 2-6 taken?

Figure 2-7   *Devils Tower. Photo by J. Steven Kite.*

Figure 2-8 *WVU Geology Capstone class at Devils Tower. Photo by J. Steven Kite.*

25. Do the ground photos in figures 2-7 and 2-8 give a better understanding of Devils Tower than you got from looking at topographic maps and aerial photos?

Google Earth for Physical Geographers

Reading assignment to be completed <u>BEFORE</u> Lab 3

Readings

Richard, Glenn A., 2019, What is Google Earth? Viewed May 2019 at http://serc.carleton.edu/sp/library/google_earth/what.html

Materials Used in Lab

- Access to a computer running Google Earth and the Internet.
- An "Intro to Google Earth for Physical Geography" website assigned in the lab syllabus, which will help navigate Google Earth using updated procedures.

Learning Objectives

After completing this exercise, students will be able to perform a series of Google Earth tasks, including:
- Identifying location coordinates (latitude/longitude) and elevations
- Navigating using place names and latitude/longitude
- Creating and using an overview map inset
- Manipulating 3D Viewer images by rotation, tilting, and zooming in or out
- Determining the role of vertical (= elevation) exaggeration in image appearance
- Using map layers, points of interest, and placemarks
- Measuring distances (both Euclidean and along a path).

What is Google Earth?

Google Earth is virtual globe software available for download on a computer, tablet, or smart phone. It maps the Earth by superimposing images obtained from satellites and aerial photos onto a 3-dimensional globe. Other layers can be added, such as political borders, roads, 3-D buildings, and personal photographs. Google Earth users can participate in this Geographic Information System by developing layers or points of interest that can be shared via the web.

Google Earth uses geographic coordinates (latitude/longitude) on the World Geodetic System of 1984 (WGS84). Google Earth uses a digital elevation model (DEM) derived from NASA Shuttle radar to allow users to find elevations and view locations in 3D.

Resolution (spatial detail) of Google Earth imagery varies across the Earth, but most images have 15 m or better resolution and most cities have 1 m or better resolution. Image dates vary, but the most recent zoomed-in views are amazingly detailed and reasonably current. Images are not taken at the same time, but are stitched together, producing changes in resolution and ground appearance as one moves from one location to the next.

Starting Out with Google Earth in Windows

Open Google Earth by double clicking in the Google Earth "globe" icon on your desktop, or selecting Google Earth from a list of programs that appears when you click on the Windows button in the lower left of the screen.

Each time you start Google Earth, the Earth appears in the main *3D viewer* window, which shows imagery, terrain, and other information about places around the globe.

- Open the *View* tab near the upper left corner to make sure the following items are selected (i.e. turned on): *Toolbar, Sidebar, Status Bar,* and *Scale Legend.* Set *Show Navigation* to *"Automatically"* or *"Always"*.

- Locate and try out the *Search Panel* in the upper right of the screen, a fill-in-the-blank tool to find places, businesses, and directions, or manage search results.

- Type in and search for your home address and other key places in your life.

Locate and try out some the following items in the tool bar along the top of the Google Earth screen, moving from left to right:

- *Hide/Show sidebar*—Click this to conceal or the display the side bar.

- *Placemark*—Click this to add a placemark pin for a location.

- *Polygon*—Click this to draw or add a polygon to the image.

- *Path*—Click this to draw a path across the image.

- *Image* **Overlay**—Click this to add a picture or image overlay on the map base.

- *Record* **a Tour**—Use this to record of a series of moves across the Earth.

- *Historical* **Imagery**—Click this clock icon to view images from other dates.

- *Sun*—Click this to show artificial sunlight and shadows across the landscape.

- *Sky*—Click this to view stars, constellations, galaxies, Mars, and Earth's moon.

- *Ruler*—Click this to measure a distances along a path or straight line.

- *Email*—Click this to email a view, image, or placemark file as an attachment.

- *Print*—Click this to print or create PDF files of views... *but don't do this in lab*!

- *View in Google Maps*—Displays the area using Google Maps in a web browser.

Go to the upper right of the Earth image to highlight the *Navigation Controls*—Use these to zoom, look, and "fly" around the Earth, and to enter *Street View*.

Locate and try out these two items along the bottom of the Google Earth screen:

- *Status Bar*—Shows image date, coordinates, elevation, and streaming status.

- *Scale Legend*—Shows an approximate scale over image

Peruse the sidebar on the left side; *"show"* the sidebar if you hid it earlier.

- *Layers panel*—Use this to display many preloaded items of interest.
- *Places panel*—Use this to locate, save, organize, and revisit Placemarks, etc.

Changing Preferences Using the View Tab

You can set a number of preferences to affect viewer imagery, as well as how icons, labels, and other elements are displayed. Some tool bar functions can also be done using the Tools or View tabs. Let's explore some changes available from the View tab.

The *Overview Map* window feature displays an inset map with a cross-hair position indicator that corresponds to the 3D view location in relation to the entire Earth.

Show the Overview Map window by doing one of the following:

- Click **View > Overview Map**, or
- Using *CTRL-M* (Holding the keyboard [ctrl] key while typing the letter M.)

1. What map projection type is used for the inset map?

2. Add a *Grid* to the 3D viewer and note the grid coordinate labels. Explain how a grid might (or might not) help you understand the map view.

3. Turn off and on the *Scale Legend*. Use the toolbar *Ruler* to measure the Scale. Do they closely agree?

4. Turn on the *Tour Guide* in the View tab; click on the small tour guide button that appeared in the lower left corner of the 3D viewer and try out a few of the tours. Generally describe what a Google Earth tour is all about.

Changing Preferences Using the Tools Tab

The Tools Tab provided another way to change the display of elements, some of which also can be changed in the View tab. Let's tour some handy changes in the *Tools* tab.

Latitude and Longitude: As you move the cursor across the image, latitude and longitude coordinates are displayed in the lower left corner. The default coordinate display is in degrees, minutes, seconds (DD°MM'SS"). If you are viewing a location in North America the "N" after the "lat" coordinate shows it is in the Northern Hemisphere, and the "W" after the "lon" coordinate shows it is in the Western Hemisphere.

5. What longitudes (rounded to the nearest degree) do all 50 United States span?

   _____ to _____

- Change lat/long coordinates units by moving the cursor to upper left corner and Clicking on *Tools > Options > 3D View,* selecting *Decimal Degrees* and clicking *OK* (which makes the Tools Tab go away) or *Apply* (which does not).

- We will leave out the Apply option from here on to avoid needless repetition.

6. Now that you selected the *Decimal Degrees setting,* which one of the coordinates, latitude or longitude, begins with a negative sign before the decimal degrees when the cursor is positioned in the coterminous "lower 48" United States?

7. Where on Earth do you have to go virtually to make the negative sign disappear?

8. To which hemisphere do you have to move the cursor to make the other coordinate become negative?

- Click on the Tools > Options > 3D View, and set *Units of Measurement* to *Meters, Kilometers* and Click *OK.*

- Note the *elev* and *eye alt* units change when you switch back from *Meters, Kilometers* to *Feet, Miles*.

- Click on the Tools > Options > 3D View, select *Use high quality terrain,* enter an *Elevation Exaggeration = 1,* and Click *OK.*

- Type *"Coopers Rock State Forest Shelter #1"* in the *Search Panel* and click *Search.* This shelter is about 7.5 miles (12 km) east of WVU Brooks Hall.

- Click and drag the image around to get a feeling for the canyon topography.

9. Explain how the look of the topography changes when you change *Terrain Elevation Exaggeration* from 1 to 3 using the Tools > Options > 3D View tab.

Navigating in Google Earth

Zooming In and Out

In the next questions, we will zoom in and out of the Grand Canyon. There are a number of ways to accomplish this. To become familiar with Google Earth, try all three:

- Zoom in using a placemark

- Zoom in and out using the mouse

- Zoom in and out using the navigation controls

Zoom in Using a Placemark

A Google Earth placemark is a visual notation that marks a location. Placemarks appear as labeled pushpins. To zoom to Grand Canyon National Park using an existing placemark:

- Locate *Sightseeing Tour* in the *Places Panel.* You may need to scroll down.
- Click on the ▶ to expand **My Places** and the **Sightseeing Tour** folder.
- Double-click the *Grand Canyon National Park* placemark.

10. Go back to the Tools tab to try out Elevation Exaggerations ranging from 0.5 to 3. What setting best helps you "see" this rugged landscape? Explain How.

Zoom In and Out with a Mouse

Once you have zoomed in using a placemark, you are ready to zoom in and out using your computer mouse. To do this:

Zoom out from the Grand Canyon by doing one of the following:

- Scroll the mouse wheel down (toward you).
- Hold down the alternate (usually the right) mouse button, drag the mouse up.

Zoom in to the Grand Canyon by doing the opposite: *i.e.* one of the following:

- Scroll the mouse wheel up (away from you).
- Hold down the alternate mouse button, drag the mouse down.

Depending on your Navigation settings, your view may tilt if you zoom in far enough.

Zoom In and Out with Navigation Controls

The navigation controls appear in the top right corner of the 3D viewer. They offer the same navigation that you can achieve with the mouse, plus some additional features.

If you have set *View > Show Navigation* to *Automatically,* the navigation controls appear whenever you move the cursor over the top right corner of the 3D viewer; they fade from sight when you move the cursor elsewhere.

Practice zooming in and out with the navigation controls. To do this:

- Mouse over the navigation controls to the vertically oriented slider.
- Zoom out by clicking the "−" zoom out button at the lower end of the slider.
- Zoom in by clicking the "+" zoom in button.

You can also use the slider or click and hold the buttons to zoom continuously in or out.

Tilting the View

Now that you can zoom in and out, you are ready to view the Earth in three dimensions by tilting your view. Tilting is a highly entertaining aspect of Google Earth, particularly when viewing hilly or mountainous terrain. As with zooming, there are multiple ways to tilt the view, using either your mouse or the navigation controls.

Tilting the View Using the Mouse

To tilt the viewpoint by zooming with your mouse:

- In the **Tools > Options > Navigation** tab, be sure **Navigation** is set to **Automatically tilt while zooming**.

- Google Earth should significantly tilt the view when you zoom in far enough, especially if you start your zoom at an **eye alt** of 35,000 feet (11,000 m) or more.

To tilt the viewpoint in a controlled fashion using your mouse:

- If your mouse has either a middle button, tilt the view by depressing the button and moving the mouse up or down.

- If your mouse has a scroll wheel, you can tilt the view by pressing the SHIFT key and scrolling DOWN to tilt the Earth into a vertical "top-down" view, or scrolling UP to tilt the Earth into a horizontal view.

Tilting the View Using On-Screen Navigation Controls

The upper circular **Look Around** joystick, which looks like a compass

- The upward pointing v-shaped "arrow" will increase the tilt of the image.

- The downward pointing arrow reduces the tilt.

- Be careful with this button, lest inadvertently you spiral totally out of control. If this happens you can recover using **View>Reset>Tilt** or **>Tilt and Compass.**

11. Change the **Elevation Exaggeration** from 1 to 3 and back. Does more or less exaggeration help you get a feel for the topography? Explain?

Street View

Google has used car-mounted cameras to take street-level views since 2007. Street-view photos are handy for anticipating what a destination looks like before one arrives.

- Click on the **Street View "Pegman"** and drag it to a location along a road or trail near the South Rim of the Grand Canyon.

12. Speculate on why Google obscures the faces of people shown in Street View.

Rotating with the Navigation Controls

The upper *Look Around* joystick can be used to rotate the view by one of two means:

- Click on the *North Direction Indicator* "N" and rotate it around the joystick.

- Click on the v-shaped arrow on the right to rotate in a clockwise direction or the arrow on the left to rotate in a counter-clockwise direction.

Remember, if you spiral out of control, you can recover via *View > Reset > Compass* or *> Tilt and Compass.*

13. Intentionally "go crazy" with the Look Around joystick for 20–30 seconds, then reset Tilt and Compass. How far did you wander from the point you started?

Moving the 3D Viewer Image across the Earth Surface

To move the 3D viewer image across the landscape using the Navigation Controls:

- Move the cursor over the navigation controls. The circular *Move Around* joystick is in the middle of the controls. It has a center hand icon.

- Click one of the v-shaped arrows to move stepwise in that direction.

- Click an arrow and hold down on the mouse button to move quicker.

You can also move using the arrow keys on your keyboard.

Fly to a Location: Latitude, Longitude

Type in this latitude and longitude in the Search panel: *27°59'18.05"N, 86°55'30.72"E.*

14. What is this feature? _____. How high is it? _____feet.

Now use the Search panel fly to a place name: *"Badwater Basin, Death Valley, CA".*

15. What are the latitude and longitude coordinates for the lowest place in North America?
_____,

16. The U.S. National Park Service and U.S. Geological Survey state the lowest point in Death Valley is 282 feet (85.5 m) below sea level. What elevation appears on the Google Earth status bar?
_____

Using Layers

Layers in Google Earth provide a variety of geographic data you can select to display over your viewing area. These includes points of interest (POIs) as well as map, road, terrain, and even building data. A list of available layers will appear in the *Layers Panel*:

You can use the layer feature of Google Earth to do a lot, including:

- Display and save points of interest.

- Display map features such as borders, roads, and terrain.

- Display 3D Buildings.

You can show or hide layers, like country, state or county borders, by checking or unchecking *Borders* within the *Border and Labels* layer in the Layers Panel.

- Turn on *Photos* in the Layers Panel; use Search to fly to *39°55'0.64"N, 80°44'41.14"W*.

17. Use Street View, tilt and rotation, or photos, to identify the cone-shaped landform?

18. Speculate on how and when it was created, and why the city where this landform is located is called the name it was given.

- Fly to: *39°38'8.44"N, 79°57'21.50"W*.

19. Where is this location? _____

- Turn on the *3D Buildings* layer and explore this area using tilt, rotation, and a range of different Elevation Exaggerations.

20. Close zooming will reveal that some larger buildings have a distinctive 3-dimensional graphic arts look and appear to rise up out of the topography. Approximately how many of these 3D buildings are displayed within 300 m (1000 feet) of this point?

- Fly to "*Eiffel Tower*" using the Search tool. Use the Navigation Tools to rotate and tilt through various views of a 3D simulation of the tower.
- Click and drag the Pegman to a street near the tower to get another perspective.

21. Which approach gives you a better appreciation of the Eiffel Tower?

Information in the Places panel can be created by anyone using Google Earth or other KML (Keyhole Markup Language) software. Most layer content is created by Google or its content partners and maintained by Google. Some layers, most notably **Photos**, are populated by users.

22. Which do you expect to have more reliably accurate locations: Google-maintained content or publicly volunteered items such as photo locations? Explain briefly.

Using Points of Interest (POIs)

This section contains tips on using points of interest:

- Locating POIs in your viewing area
- Saving or copying POIs to *My Places*

- Viewing Layer subcategories

- Tuning the display of POIs

Fine Tuning POI Displays

Because Google Earth can deliver hundreds of POIs to a single view, the icons marking them may be modified to make viewing easier.

Common issues in POI display include:

- A POI doesn't appear over an area where one is expected

- No icons appear at all

- Icons overlap each other, making it hard to see the one being covered

Try the following to resolve the issue:

- Zoom in closer to see if icons appears. Google Earth Icons, like Google Maps symbols, appear at different scales (Google calls them "elevations"), and not all icons appear from high elevation. In addition, zooming to a finer scale often resolves the problem of icons that overlap when viewed from a coarser scale.

- Check the **streaming meter** to make sure downloading is complete. This circular meter appears at the bottom right of the 3D viewer. If you are sufficiently zoomed in and no POI appears, then imagery, terrain, and other data still may be being streamed to your computer from the Google Earth servers.

- Adjust icon size from medium to small or large, depending upon whether you typically view the Earth from a higher elevation or a lower elevation. Do this by changing the selecting *Tools* > **Options** > *Labels/Icon size* setting.

Measuring Distances Using the Ruler

Google Earth has tools to measure distances along a line or path. Distances are calculated using coordinates from point to point. Results are given as both straight-line *Map Length* that does not consider topography, and *Ground Length* that factors in elevation changes. To use the ruler:

- Position the image you want to measure within the 3D viewer.

- Select **Ruler** from the Toolbar or Tools menu and a Ruler dialog box appears.

- Move the dialog box to a region of your screen that doesn't obstruct the 3D viewer or get in the way of your measurement.

- Choose **Line** or **Path** and the units of measure.

- Click in the 3D viewer to set the beginning point for your line or path.

- Click on the endpoint of the line or continue clicking until a whole path is traced.

- A red dot indicates the beginning point of your shape, and a yellow line connects to it as you move the mouse.

- The total measurement of the line or path is shown in the Ruler dialog box, but note you can choose other measurement units after measuring.

23. Measure the straight-line (Euclidean) distance between the Beechurst PRT station and Monongalia County Ballpark in Granville, WV. _____mi.

24. Now measure the path you would drive between the two points: _____mi.

25. How long is the longest runway at the Morgantown Airport? _____ft.

Measuring Distances Using Google Tools

Google Earth includes measurement functions familiar to GPS navigation users.

- Use the *Get Directions* option in the Search Panel to determine the driving distance from the WVU Ruby Memorial Hospital in Morgantown to Madison, WV.

26. How far is the drive and how long would it take, using the fastest route available? _____mi. _____ minutes.

27. Use the Ruler tool to measure the Euclidean distance that a 150 mile/hour life flight helicopter might fly from Madison to Ruby Memorial Hospital. How far is the flight? _____mi. How long would it take? _____ minutes.

Exploring the Scale of Appalachian Mountain Top Mining

Follow these steps to view a layer focused on Appalachian mountaintop mining.

- Double-click on the Layers folder to expand it.

- Expand the *Global Awareness* folder within the Layers folder to view a list of Global Awareness KML files.

- Check the box and double-click on *Appalachian Mountaintop Removal*.

- You may want to uncheck other layers to see these symbols more clearly, but make sure **Borders and Labels** stay checked.

28. List the four states with POIs in the Appalachian Mountaintop Removal layer?

Three types of icons should appear. One is a flag in a circle, another is a triangle with a bull's eye, and the third is a yellow hard hat.

- Click on a few of each.

29. Who created the Appalachian Mountaintop Removal layer and for what purpose?

- Click on the icon for the *User's Guide*, (which may appear near Knoxville, Tennessee.) Scroll down to the *High-Resolution Mine Tour* and click the link to open the tour.
- Examine the Hobet Mine Complex in the Appalachian Mountaintop Removal layer.
- Reset Tilt and Compass.
- **Use the Historical Imagery slider to look at changes in the extent of mining at the Hobet Mine area?**

30. **When did** the extent of active and reclaimed surface **mining grow beyond** the yellow mapped mine boundaries shown in the **Appalachian Mountaintop Removal** KML layer.

31. What is the longest dimension of the Hobet Complex? _____(mi)

32. What is the distance across urban Morgantown? _____ (mi)

Geography Treasure Hunt

Work in teams of several classmates and time how long it takes to correctly complete this page. There will be a 1 minute penalty for every wrong answer!

Answer these brief questions

1) How long is the world's longest bridge over water, which connects Metairie and Mandeville, Louisiana? _____

2) Give elevation of Mount Logan, the highest peak in Canada and second highest peak in North America? _____

3) What are the latitude and longitude of the southernmost bare ground in the U.S.? (*E makaala: the answer may not lie where you expect.*) _____

4) How far is it between the lowest point in the lower 48 United States and the highest point in the lower 48 states? _____

5) Give the latitude and longitude of McMurdo Station, Antarctica, the southernmost bare ground accessible by ship? _____

What is located at these coordinates?

Work in teams to find these location. Give place name and brief description of each.

6) 25° 20' 39" S, 131° 02' 04" E

7) 29° 58' 44" N, 31° 08' 05" E

8) 37° 44' 36" N, 119° 32' 06" W

9) 45° 58' 34" N, 7° 39' 24" E

10) 13° 09' 47" S, 72° 32' 44" W

Earth-Sun Relationships

Reading and Watching assignments to be completed <u>BEFORE</u> Lab 4

Readings

Strahler, A.N. **<u>Introducing Physical Geography</u>**, John Wiley & Sons, New York, **Chapters 1, 2 & 3**.
Kaplan, Rebecca, 2013, **<u>Reasons for the Seasons</u>** (5:20 long video), retrieved May 2019 from https://www.youtube.com/watch?v=DD_8Jm5pTLk.
Kurdistan Planetarium, 2011, **<u>Mechanism of the Seasons</u>** (5:59 long video), retrieved May 2019 from https://www.youtube.com/watch?v=WLRA87TKXLM.

Materials Used in Lab

- 12 inch globe
- Rand McNally's Goode's World Atlas

Learning Objectives

Upon successful completion of this lab, students will be able to:

- Identify the subsolar point during equinoxes and solstices
- Understand the relationship between latitude and temperature
- Generate a hypothesis about causes of global temperature variation

Earth-Sun Relationships

1. Draw arrows on Figure 4-1 to show the direction Earth revolves around the Sun. Label
   - N (for north) and S (for south) poles
   - Northern Hemisphere winter and summer *solstices*
   - Northern Hemisphere vernal and autumnal *equinoxes*
   - Draw and label the Equator (Eq) on all four Earth positions.

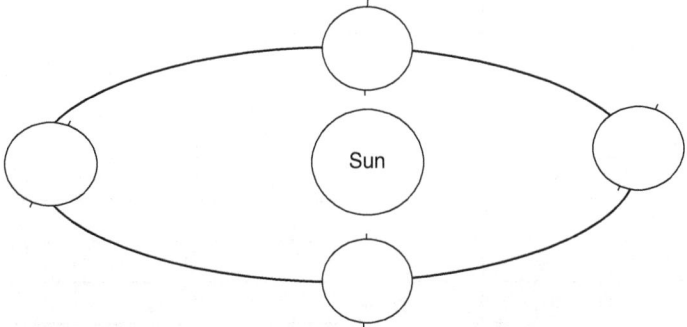

Figure 4-1 *Seasonal Earth-Sun relationships.*

2. At what latitude on the Earth's surface will the Sun's rays be vertical (directly overhead at noon) during the Northern Hemisphere summer solstice?

3. At what latitude will the Sun's rays be vertical at noon during the Northern Hemisphere winter solstice?

4. During the vernal and autumnal equinoxes in the northern hemisphere, at what latitudinal position on the Earth's surface will the Sun's rays be directly overhead at noon?

5. Sketch arrows representing the direction of the Sun's incoming rays during each solstice and each equinox on figure 4-1.

The Analemma: Are All Days 24 Hours Long?

Table 4-1 shows sunrises and sunsets in Morgantown, WV, for selected days 2019–2020, from https://www. timeanddate.com. A 24 hour clock was used in order to simplify the math.

Table 4-1. Selected sunrises and sunsets, Morgantown, WV, 2019–2020.

| Date | Eastern Daylight Time | | Eastern Standard Time | | Length of Day |
	Sunrise	Sunset	Sunrise	Sunset	
June 21	5:52	20:50	4:52	19:50	
July 21	6:09	20:42	5:09	19:42	14:33
Aug 22	6:38	20:06	5:38	19:06	13:28
Sept 23	7:08	19:15	6:08	18:15	
Oct 21	7:35	18:32	6:35	17:32	10:57
Nov 21	-	-	7:10	17:00	9:50
Dec 7	-	-	7:26	16:55	9:29
Dec 14	-	-	7:32	16:56	9:24
Dec 21	-	-	7:36	16:58	
Dec 28 '10	-	-	7:39	17:03	9:24
Jan 4 '11	-	-	7:40	17:08	9:28
Jan 21	-	-	7:36	17:26	9:50
Feb 21	-	-	7:04	18:02	10:58
Mar 19	7:24	19:31	6:24	18:31	
Apr 20	6:34	20:03	5:34	19:03	13:29
May 21	6:00	20:33	5:00	19:32	14:33

6. **Lengths of daylight** have been provided for most dates, but you must subtract times to calculate it for the four dates left off the chart: the two solstices and the two equinoxes; *enter these lengths of day into Table 4-1.*

Equinoxes appeared to be dates with nights and days of equal length to ancients measuring time with a sundial, but your calculations show daylight is not exactly 12 hours long during an equinox. The atmosphere bends the Sun's rays so that the Sun appears a few minutes longer each day than it would if the Earth had no atmosphere.

Have you noticed that the earliest sunset occurs a week or so before the shortest day of the year: the winter solstice? The latest sunrise is well after the solstice, which is part of why it is so hard to get up for morning classes in January. This strange offset between the earliest sunset, the winter solstice, and the latest sunrise stems from the fact that the speed at which the Earth orbits around the Sun is not constant over the year.

The rate of Earth's rotation relates to its distance from the Sun: the closer the Earth is to the Sun, the faster it orbits around the Sun. Day length over a year averages 24 hours. From an astronomical perspective of a day being the time between the middle of one night to the middle of the next night, some days are actually up to 40 seconds less or 20 seconds more than 24 hours long!

7. What is the date of Earth's **perihelion** this year, when the Earth is closest to the Sun? In which Northern Hemisphere season does this date occur?

8. Does the distance of the Sun from the Earth control the seasons? Explain your answer.

An *analemma* shows the pattern the Sun makes if you photograph the sky at exactly the same minute on many dates through the year.

Two NASA Astronomy pictures of the day may help you grasp the analemma concept:

- Apollo's Analemma from Greece is at http://apod.nasa.gov/apod/ap130922.html

- Analemma over New Jersey links to a 49 second time lapse-sequence at http://apod.nasa.gov/apod/ap071204.html

Many analemma on-line videos show how an analemma is created, including these two:

- The seasonal progression of the analemma is captured in nicely produced 2 minute 18 second video by Luca Vanzella: 2014, The Solar Analemma over Edmonton, retrieved 7 June 2017 from https://www.youtube.com/watch?v=jQT5XRdrqvw

- Aryan Navabi, of the the Kurdistan Planetarium, has an highly informative 10 minute video at https://www.youtube.com/watch?v=82p-DYgGFjI, but it goes into some topics a more deeply than we need for this lab exercise.

The analemma in figure 4-2 shows the position of the Sun at 12 noon throughout the year at Washington, DC. It was published by Marc A. Murison of the United States Naval Observatory Astronomical, and was downloaded in 2004 from USNO web site, but sadly the URL: http://arnold.usno.navy.mil/murison/SunAltAz/ WashDC1997.gif is now a dead link.

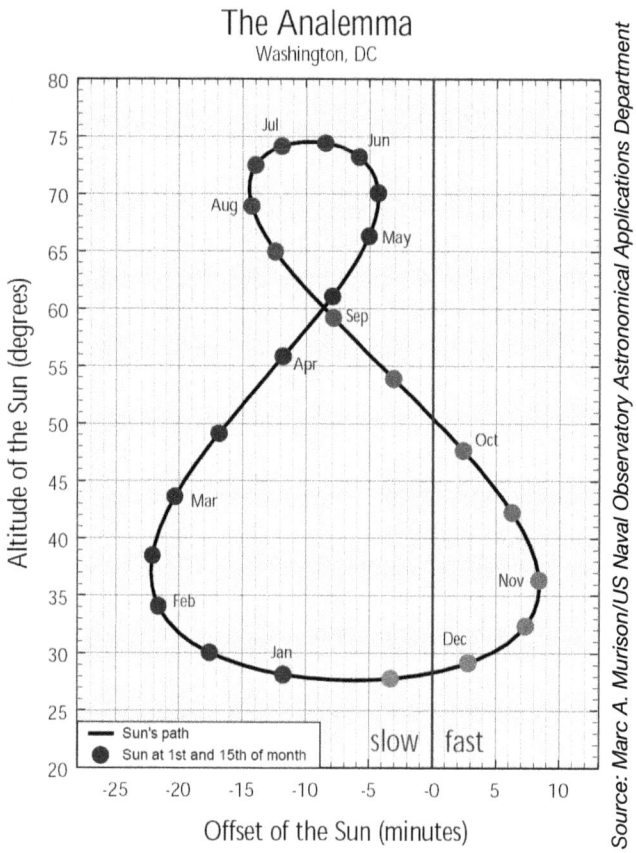

Figure 4-2 *Analemma showing the Sun's position at noon over a year at Washington, DC.*

Note the Sun reaches the highest point in its arc across the sky as much as 8 minutes fast (11:52 a.m.) or as much as 22 minutes slow (12:22 p.m.). The greatest rate of change coincides with ***perihelion*** in early January, when the Earth is only 147,100,000 km (91.4 million miles) from the Sun. The slowest rate of change is during ***aphelion***, when the Earth is farthest from the Sun, 152,100,000 km (94.5 million miles).

9. The figure 4-2 analemma shows the angle of the Sun from the horizon, labeled as "altitude". What is the lowest angle above the horizon for the Sun at noon in Washington in winter?

10. What is the highest angle for the Sun at noon in Washington, DC, in summer?

11. What is the angle above the horizon for the Sun at the equinoxes?

12. You can calculate the latitude of a location by subtracting the maximum Sun angle above the horizon at the equinox from 90°. Subtract your answer to question 11 from 90° to find the latitude of Washington, DC.

13. Confirm your answer to the previous question by looking up the latitude of Washington on a map. Did your two values of latitude agree?

Global Temperatures

14. Look at the "Relationship between net radiation and temperature" figure in chapter 3 of the Strahler textbook. Focus on the annual temperature cycles for Hamburg and Yakutsk, two cities with similar net radiation curves. What are the average July C° temperatures for these places? What are the average January C° temperatures?

15. Look at the world population density map in your atlas. What is the approximate population density for the regions where these two cities are located?

16. Do you think the annual temperature cycle at these two locations have anything to do with the population density? Explain the basis for your answer.

17. What are the average July and January C° temperatures in Manaus, Brazil?

18. How does the population density for the area near Manaus compare to Hamburg?

19. Is the annual temperature cycle the only control of population density? What other physical geography factors are likely to be important?

Use the "Maritime and continental annual air temperature cycles" figure in chapter 3 of the Strahler textbook to compare annual temperature cycles for two stations at the same 50°N latitude. Winnipeg, Manitoba, Canada, has a continental climate, whereas the Scilly Islands, southwest of Cornwall, England, have a maritime climate.

20. How does the surrounding Atlantic Ocean influence the range of temperatures and the timing of the maximum and minimum temperatures on the Scilly Islands?

Examine both the world temperature maps in a world atlas and the mean monthly air temperatures figures in the Strahler textbook to answer the following questions.

21. Generally, which regions experience the greatest range in temperatures in a year?

22. Which specific region has the greatest temperature range each year?

23. Which region in North America has the greatest temperature range each year?

24. Generally, which regions experience the least temperature variation through the year?

25. Which small state in the USA has the least temperature range in a typical year?

26. How does the annual variation in Western Europe compare to that at the same latitude on the east coast of North America?

27. Which has a warmer climate: northern Norway or central Greenland? Make one or two hypotheses on why there is such a pronounced difference.

28. Use the polar projection maps to identify the coldest place on Earth in January. In July.

29. What is the mean July temperature at the North Pole?

30. What is the mean January temperature at the South Pole?

31. Remembering that seasons are reversed in the two hemispheres, which has the coldest winter temperatures: the North Pole or South Pole?

32. What explanations can you and your classmates give for the variations in global temperatures?

Lab Exercise 5

Temperature and Lapse Rates

Reading assignments to be completed <u>BEFORE</u> Lab 5

Readings

Strahler, A.N., **Introducing Physical Geography**, John Wiley & Sons, New York. **Chapter 2.**
Peterman, Erin, 2009, A Cloud with a Silver Lining: The Killer Smog in Donora, 1948, http://
pabook2.libraries.psu.edu/palitmap/DonoraSmog.html (viewed May 2019).

Materials Used in Lab

- Rand McNally's Goode's World Atlas
- Hand calculator

Learning Objectives

Upon successful completion of this lab, students will be able to:
- Create X-Y graphs from quantitative data
- Read and interpret X-Y graphs
- Convert temperatures between Fahrenheit and Celsius scales
- Plot annual temperature curves for different climates
- Calculate an observed lapse rate
- Understand the causes and nature of temperature inversion

Temperature Scales

Temperature has been a key environmental factor since before the dawn of civilization, but a factor difficult to quantify until the invention of the thermometer. The first modern thermometer was developed by Daniel Gabriel Fahrenheit in 1714. His *Fahrenheit scale* for measuring temperatures (designated by °F) still remains in use in the United States and six small island nations, each with a population less than 500,000.

The other 95+ percent of the world measures temperature using a scale developed by Anders Celsius and improved by Carolus Linnaeus in the 1740s. The *Celsius scale* (°C) is closely tied to fundamental physical phenomena that occur on the Earth surface at level: 0 °C being the temperature at which water freezes into ice and 100 °C the point at which water boils away as water vapor (steam). The 100 C° difference between the phases of water on the Celsius scale has led to an informal alias of *Centigrade scale*.

Celsius degree units are related to the *Kelvin scale* (°K), in which 0 °K is *absolute zero*—as cold as anything can be. Although very handy in physics, astronomy, and planetary geology, the Kelvin scale is less useful for everyday Earth surface observations.

One temperature semantic is worth noting. Temperature readings should be stated as "degrees Celsius" (°C) or "degrees Fahrenheit" (°F), but temperature differences are expressed as "Celsius degrees" (C°) or "Fahrenheit degrees" (F°). Exact phrasing may not be critical most of the time, but communicating like you "know what you are talking about" creates a good impression at a job interview or in the workplace!

Thinking in Celsius and Fahrenheit

The more one uses a scale to measure or discuss temperatures, the more it makes sense. As in learning a new language, to truly understand the Celsius temperature scale most Americans need to learn to think in Celsius terms. International students may need to learn to think in Fahrenheit. Table 5-1 may help you internalize the two scales.

1. Fill in the 2nd and 3rd columns to show what you typically would wear in Morgantown during the afternoon and late nights in January, April, July and October?

Table 5-1. Seasonal clothing for Celsius and Fahrenheit temperatures, 1981–2010 climate normal, from www.usclimatedata.com/climate/morgantown/west-virginia/united-states/uswv0507 (retrieved May 2019).

Month	Afternoon Wear	Late Night Wear	Maximum	Minimum
January			4 °C (39 °F)	–6 °C (21 °F)
April			18 °C (64 °F)	4 °C (39 °F)
July			28 °C (83 °F)	17 °C (63 °F)
October			18 °C (65 °F)	6 °C (43 °F)

Temperature Scale Conversions

No matter which scale is most familiar, the ideal situation for world travelers and anyone involved in the global economy (i.e. everyone) is to "understand" the physical meaning of both Celsius and Fahrenheit temperatures. However, there may be situations in which you need to numerically convert from one scale to the other.

The equation to **convert Celsius temperature (°C) to Fahrenheit (°F)** is as follows:

$$°F = (9/5 \ °C) + 32)$$

i.e. divide the Celsius temperature by 5, multiply by 9, and add 32.

The equation to convert **Fahrenheit temperature (°F) to Celsius (°C)** is as follows:

$$°C = 5/9 \ (°F - 32)$$

i.e. subtract 32 from the Fahrenheit temperature, divide by 9, and multiply by 5.

2. Use the two temperature conversion equations to <u>fill in the six data points missing from the next three temperature tables</u>. Either the °**F** or °**C** temperature is given in each case, but you will have to write the temperature in the other scale in the open cell. Precipitation data are is also provided, but we will not use the precipitation data until a later lab exercise, when we will construct **climographs**.

Table 5-2. Whittier, Alaska: NOAA National Weather Service Monthly Averages 1971–2000, from Weather.com. Whittier is one of the wettest locations in the United States.

	Jan	Feb	Mar	Apr	May	Jun	Jul	Aug	Sep	Oct	Nov	Dec
Mean Temp °F		29	33	39	46	53		56	49	39	32	29
Mean Temp °C	−3	−2	1	4	8	12	14	13	9	4	0	−2
Mean Precip (in)	18.96	13.84	13.35	14.28	12.91	9.63	9.18	13.73	21.89	20.00	16.51	20.94
Mean Precip (cm)	48.16	35.15	33.91	36.27	32.79	24.46	23.32	34.87	55.60	50.80	41.94	53.18

Table 5-3. International Falls, Minnesota: NOAA National Weather Service Monthly Averages 1971–2000, from Weather.com. The northernmost town in the "Lower 48", International Falls is frequently the coldest point in the whole country during winter.

	Jan	Feb	Mar	Apr	May	Jun	Jul	Aug	Sep	Oct	Nov	Dec
Mean Temp °F		11	24	39	53	62		64	53	42	24	9
Mean Temp °C	−16	−12	−4	4	12	17	19	18	12	6	−4	−13
Mean Precip (in)	0.84	0.64	0.96	1.38	2.55	3.98	3.37	3.14	3.03	1.98	1.36	0.70
Mean Precip (cm)	2.13	1.63	2.44	3.50	6.48	10.11	8.56	7.98	7.70	5.03	3.45	1.78

Table 5-4. Death Valley, California: NOAA National Weather Service Monthly Averages 1971–2000, from Weather.com. Death Valley is 86 m (282 ft) below sea level in the Mojave Desert, as you discovered in an earlier exercise!

	Jan	Feb	Mar	Apr	May	Jun	Jul	Aug	Sep	Oct	Nov	Dec
Mean Temp °F	52	60	67	76	85	95	101	99	90	77	61	51
Mean Temp °C	11	16	19	24	29	35		37	32	25	16	
Mean Precip (in)	0.35	0.42	0.42	0.12	0.10	0.05	0.11	0.14	0.19	0.13	0.12	0.18
Mean Precip (cm)	0.89	1.07	1.07	0.30	0.25	0.13	0.28	0.36	0.48	0.33	0.30	0.46

Graphing Temperatures

3. Plot the monthly mean temperatures for Whittier, International Falls and Death Valley on the graph below. Use different colors for each station and connect the points throughout the year for each station with a line of the same color.

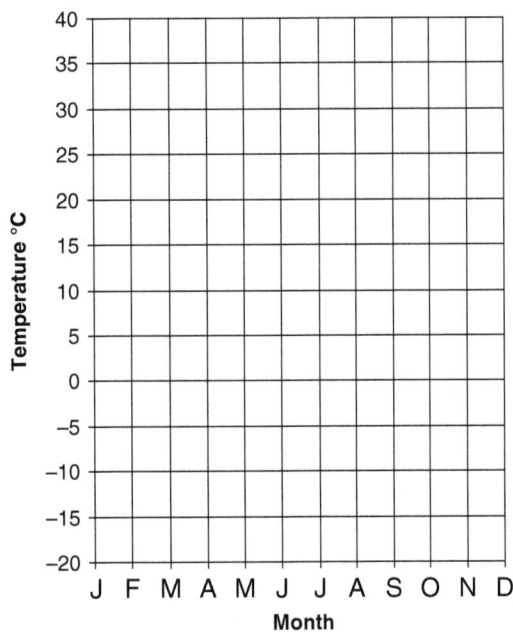

4. Which station is coldest in January?

5. Which station is coldest in July?

6. Which station has the greatest temperature range?

Mean monthly temperatures for six of the following seven stations are plotted in Figure 5-1, following tables 5-11. The climates at these stations are less extreme, but the data give a sense of the temperature variation in the eastern USA. All data are NOAA National Weather Service Monthly Averages, retrieved from Weather.com in 2004.

Table 5-5. Pittsburgh, Pennsylvania:

	Jan	Feb	Mar	Apr	May	Jun	Jul	Aug	Sep	Oct	Nov	Dec
Mean T °F	27	31	40	50	60	68	73	71	64	53	42	33
Mean T °C	−3	−1	4	10	16	20	23	22	18	12	6	1

Table 5-6. Charleston, West Virginia:

	Jan	Feb	Mar	Apr	May	Jun	Jul	Aug	Sep	Oct	Nov	Dec
Mean T °F	33	37	45	54	62	70	74	73	66	55	46	38
Mean T °C	1	3	7	12	17	21	23	23	19	13	8	3

Table 5-7. Canaan Valley, West Virginia:

	Jan	Feb	Mar	Apr	May	Jun	Jul	Aug	Sep	Oct	Nov	Dec
Mean T °F	24	27	35	44	54	61	65	63	58	48	38	29
Mean T °C	−4	−3	2	7	12	16	18	17	14	9	3	−2

Table 5-8. Washington, DC:

	Jan	Feb	Mar	Apr	May	Jun	Jul	Aug	Sep	Oct	Nov	Dec
Mean T °F	33	36	44	54	64	73	78	76	69	57	47	38
Mean T °C	1	2	7	12	18	23	26	24	21	14	8	3

Table 5-9. New York, New York:

	Jan	Feb	Mar	Apr	May	Jun	Jul	Aug	Sep	Oct	Nov	Dec
Mean T °F	33	35	43	52	63	72	77	76	69	58	48	38
Mean T °C	1	2	6	11	17	22	25	24	21	14	9	3

Table 5-10. Myrtle Beach, South Carolina:

	Jan	Feb	Mar	Apr	May	Jun	Jul	Aug	Sep	Oct	Nov	Dec
Mean T °F	46	49	56	63	71	77	81	80	75	64	56	48
Mean T °C	8	9	13	17	22	25	27	27	24	18	13	9

7. What is the difference between temperatures at Myrtle Beach and Canaan Valley in January? In July?

Table 5-11. Morgantown, West Virginia:

	Jan	**Feb**	**Mar**	**Apr**	**May**	**Jun**	**Jul**	**Aug**	**Sep**	**Oct**	**Nov**	**Dec**
Mean T °F	30	33	42	52	61	69	73	72	65	54	44	35
Mean T °C	−1	1	6	11	16	21	23	22	18	12	7	2

8. Plot the monthly mean temperatures for Morgantown on figure 5-1 below.

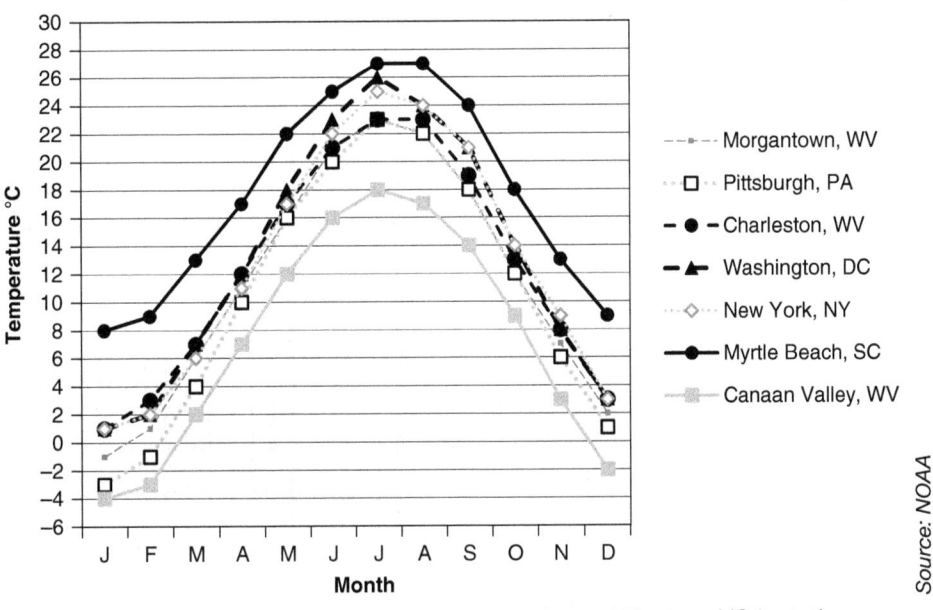

Figure 5-1 *Monthly mean temperatures at selected Eastern USA stations.*

Source: NOAA

Interpreting Temperature Graphs

9. Which months are likely to see long-lasting snow accumulations in Canaan Valley? In Pittsburgh? In Washington? In Myrtle Beach?

10. Were differences between New York and Washington as you expected? Explain.

11. Use a world atlas to find which is further north: New York or Pittsburgh?

12. What temperature-controlling factors other than latitude may be important in the eastern US?

Atmospheric Temperatures and Lapse Rates

13. The lower atmosphere or *troposphere* is heated from the ground up. What type of radiation is most important in regulating troposphere temperatures? Where is this radiation source?

14. Graph the temperatures and altitudes in the following table to create a *temperature profile* on the following grid:

Altitude (km)	Temp °C
0	35
2	22
4	9
6	−4
8	−17
10	−30
12	−43
14	−54
16	−52
18	−50
20	−49

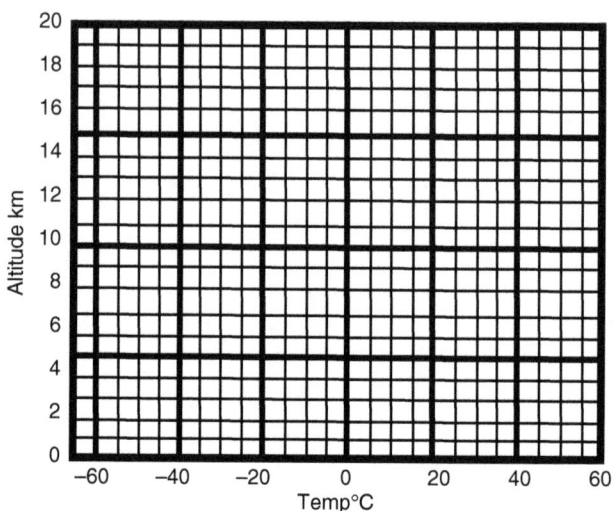

15. Calculate the average *lapse rate* (change in temperature with change in altitude) between 0 and 14 km, based on the data you plotted on the previous graph.

16. Given the temperature at sea level, what season does this profile likely represent?

17. At what altitude in this temperature profile would water droplets freeze?

18. If the sky in which this temperature profile was measured has clouds topping out at 12 km (40,000 ft) tall, are these clouds made of water vapor or ice crystals?

19. If the surface temperature is a chilly 2 °C and the lapse rate is −6 °C/km, predict the °C temperature at an altitude of 6 km (20,000 ft)? Convert this °C temperature to °F.

20. As added perspective, imagine World War I aviators flying in open cockpit aircraft over northern France at altitudes up to 6 km. What temperatures did they experience on winter flights?

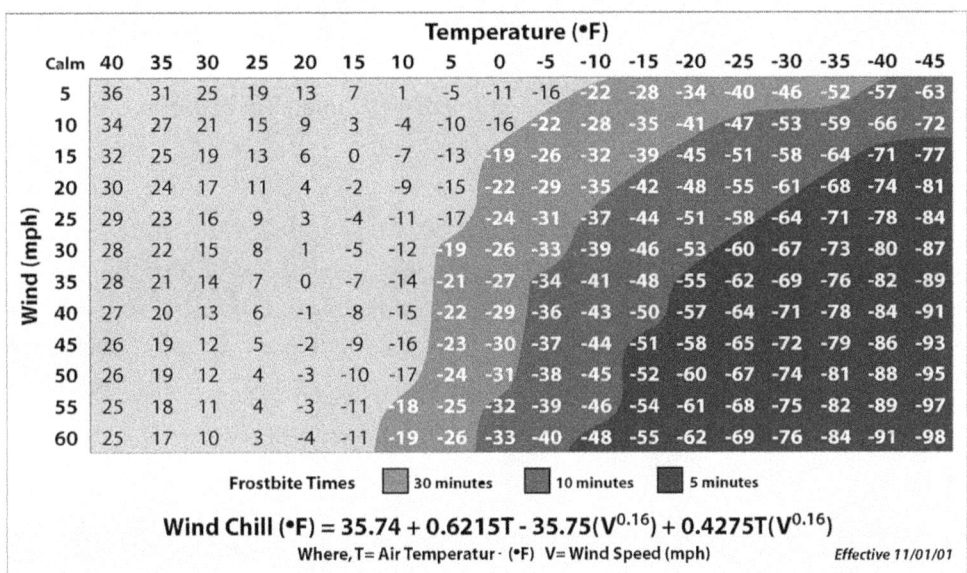

Figure 5-2 NOAA National Weather Service Wind Chill Chart, Effective 2001. https://www.weather.gov/safety/cold-wind-chill-chart (retrieved June 2018).

21. Many World War I aircraft reached speeds of 120 miles/hour (mph). If small windshields reduced the flow of air past an aviator's head by half to 60 mph, calculate the *wind-chill factor* in °F using figure 5-2 and your answers to 19 and 20.

22. Canaan Valley has an altitude of 1000 m above sea level. Morgantown has an altitude of 300 m; Washington, DC, an altitude of ~ 0 m. Looking back at the temperature data on tables 5.5–5.10, compare the monthly mean temperatures of Canaan Valley, Morgantown, and Washington. Using the lapse rate to calculate the expected temperatures and compare them to the temperatures in the tables. Are the monthly mean temps what you would expect?

23. At 1482 m (4863 ft) above sea level, the highest point in West Virginia is Spruce Knob, not far from Canaan Valley. Round 1482 m up to 1500 m to make the math simple and use a 6 °C/1000 m lapse rate to calculate expected mean January and July temperatures on Spruce Knob. Refer to table 5.7.

24. Whittier, AK, (Table 5-2) is at sea level, but mountains surrounding the town exceed 1000 m in elevation. Assuming a 6 °C/1000 m lapse rate, determine how many months have average temperatures below freezing (0 °C) on top of these mountains.

25. In what form does most of the precipitation fall on the mountains above Whittier?

All precipitation data for this lab are given in *water equivalents*... *i.e.* all snow, sleet, *etc.* are reduced to the amount of water left when they melt. Water equivalence of snowfall varies (mostly with air temperature) and the ratio of snow to water equivalence typically ranges from 5 to 20, but use 10 cm of snow = 1 cm of water for this exercise.

26. What is the total (water equivalent) precipitation at Whittier from October to April?

27. Approximately how many meters of snow fall on mountains above Whittier yearly?

Temperature Inversions and the 1948 Donora Disaster

28. What is a *temperature inversion*?

29. Describe the 1948 disaster in Donora, Pennsylvania, based on Peterman (2009) reading—http://pabook2.libraries.psu.edu/palitmap/DonoraSmog.html:

30. Find Donora in an atlas or map app. How far is Donora from your campus?

31. Use the chart below to plot a temperature profile of a temperature inversion. Compare this to the plot you made for question 14 and explain how inversions differ from normal conditions?

Altitude (km)	Temp °C
0	−2
2	1
4	4
6	−3
8	−14
10	−28
12	−42
14	−55
16	−54
18	−52
20	−50

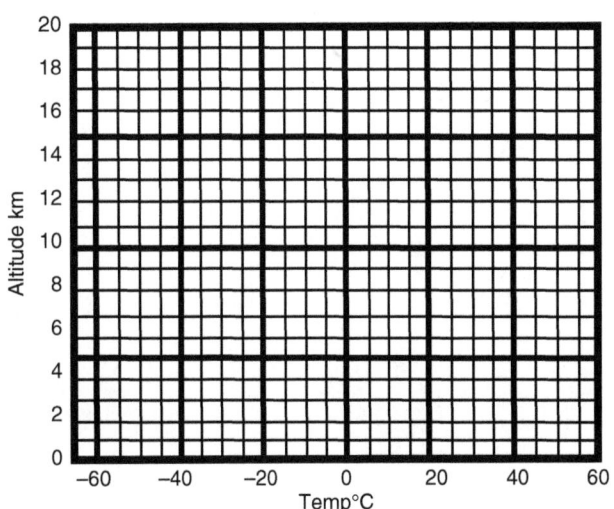

32. Given what you know about the movement of warm and cool air, how would pollution from a smokestack 300 m tall move in the atmosphere illustrated in the temperature inversion graph above? Draw on the graph where pollution might get trapped.

33. What physical similarities exist between the topography of Morgantown, West Virginia, and that of Donora, Pennsylvania?

34. How and when might temperature inversions create public health issues if fossil-fuel burning electricity-generating power plants or industrial facilities around Morgantown give off unhealthy emissions?

Atmospheric Moisture

Readings

Strahler, **Introducing Physical Geography**, John Wiley & Sons, New York. **Chapter 4.**

Materials

Raven maps of the "Lower 48" and West Virginia (Alan Cartography, 1992, 1993), World Atlas, Cloud Charts; a sky, preferably with clouds, provided by nature.

Learning Objectives

Upon successful completion of this lab, students will be able to:

- Identify simple cloud types
- Interpret the atmospheric processes associated with different clouds types
- Make short-term weather forecasts based on cloud identifications
- Understand how mountains/elevation affect precipitation averages
- Identify locations affected by orographic precipitation on elevation maps

Clouds

1. Work with lab colleagues to reach a consensus on the definition of the term **cloud**:

The National Oceanic & Atmospheric Administration-Weather Service (NOAA-NWS) glossary http://w1.weather.gov/glossary/index.php [viewed May 2019] defines **cloud** as "A visible aggregate of minute water droplets or ice particles in the atmosphere above the Earth's surface".

2. How did your group's personal definition differ from the NOAA-NWS definition?

Cloud charts (e.g. http://scool.larc.nasa.gov/pdf/1-PageCloudChart/Cloud_ID.pdf [viewed May 2019]) show clouds can be subdivided according to their altitude above the surface:

Low clouds occur at 0 to 2 km (0–6,500 ft),
Mid-level clouds occur at 2 to 7 km (6,500–23,000 ft),
High clouds occur at 7 to 15 km (23,000–45,000 ft), and
Convective clouds span two or more levels, perhaps ranging from 0 to 18 km (0 to 60,000 ft),

Although there are many different clouds types, weather watchers can appreciate most of their diversity by understanding five Latin roots: *stratus, cumulus, cirrus, alto,* and *nimbus*. The first three terms can be used alone, but many cloud types are named by combining these Latin roots.

The typical low level cloud is a gray cloud layer with a fairly flat uniform base called *__stratus__*.

The simplest convective (vertically developing) clouds are called *"__cumulus__"*: dense, white, detached clouds, with sharp outlines. *Cumulus* clouds typically have horizontal bases, but show upward growth in the form of domes, mounds, or towers.

Mid-level clouds are designated by the prefix *"__alto__"*, as in *__altostratus__* or *__altocumulus__*.

The typical high level cloud is composed of ice crystals, appearing in the form of delicate white to semi-transparent filaments, patches, or narrow bands, called *__cirrus__*.

The Latin word *"__nimbus__"* translates into "dark cloud", indicative of significant precipitation.

Obviously clear skies bring "fair" weather, but some cloud types are also usually associated with nice weather, although they may be harbingers of precipitation in the near future.

3. Look at the cloud charts and use your lab group's personal experience to list three cloud types likely to be in the sky without causing rain at the moment of observation.

4. Which fair-weather clouds are likely to give "bumpy" airplane trips because of the turbulence that accompanies convection?

5. Which cloud types would make you inclined to carry an umbrella or wear a rain coat?

6. Which one of these foul-weather clouds are most likely to give frightening, nauseating airplane trips with extreme turbulence and rapid convection, unless your pilot wisely avoids them.

7. Which of the foul-weather clouds are likely to give boring, uneventful airplane trips with little turbulence and no views of the earth's surface except during takeoff and landing.

8. What is a cloud at the earth's surface called? (Hint: it rhymes with dog, hog, and log).

9. Some cloud charts dub **contrails** (condensation trails formed from moisture in airplane exhaust) a separate cloud type, but if your group were forced to place them in one of the Latin cloud types, which one would you pick?

10. Re-examine the concept of environmental lapse rates and your plot of temperatures with changing altitude in lab 5, and come up with a consensus of the members in your lab group over at what altitudes cloud particles will be water droplets vs. ice crystals during 35°C (~95°F) summer afternoons.

11. At what altitudes will cloud particles will be ice crystals on a –10°C (~14°F) winter day.

12. In the space below, use graphite gray or colored pencils to draw and label four different cloud types of your choice.

13. Review the clouds shown in cloud charts and fill out the following table:

Fair-Weather Clouds	It is going to rain within the next day clouds	It is going to rain soon if it isn't raining already clouds	Time to get inside ASAP clouds

Ten Minute Field Trip

Scientists make sketches to complement field or lab notes. Some create wonderful art, while others fall short of creating something visually pleasing. No matter how bad the artist, the process of studying an object, analyzing that object, and trying to relate it to others (or to personal notes for future reference) is a great learning experience. Question 14 asks you draw clouds, not for sake of creating art, but to help you understand what each cloud types is all about.

14. Go outside or to a window to observe, identify, and sketch today's clouds in the space below. Note their apparent level and use cloud charts, textbooks, and online resources to identify them. Cloud charts cannot leave the lab, and many textbooks provide few good cloud photographs, but, for cloud identification help, check out NASA's S'COOL: On-Line Cloud Chart at http://scool.larc.nasa.gov/printables-guides-CloudChart.html (viewed May 2019).

Where Not to Live if You Hate Rain!

The orographic effects on rainfall in the Western United States are dramatic and widely known (see chapter 4 in Strahler), but there are distinctive orographic patterns in the Appalachians that are less appreciated by the public and some geographers. The next part of this lab will focus on how rainfall is distributed across the central Appalachians.

In the mid-latitudes the dominant circulation caries "weather" from west to east, so many air masses that originate in the Pacific Ocean pass over the Appalachians. However, passage over the Cordilleran Mountains in the western U.S. generally "wrings" moisture out of Pacific air, and the Gulf of Mexico is the most important source of moisture for precipitation in the eastern two-thirds of the "Lower 48" and much of Canada. Except during certain events, Atlantic Ocean moisture is dominant only along the East Coast.

Both Gulf and Atlantic moisture are critical to the Central Appalachians, but points west of the Blue Ridge Mountains are dominated by moisture from the Gulf of Mexico. Subtle differences in **aspect** (the direction a slope faces) and topography can make a big difference in rainfall and snow accumulation, and a noticeable difference in the amount of cloudiness.

15. Study the Raven map of the conterminous United States, and predict where rain shadows are likely to occur throughout the western "Lower 48 states."

16. Predict areas the western "Lower 48 states" that are likely to receive excessive precipitation because of orographic effects.

17. Check your answer to questions 15 and 16 to the Average Annual Precipitation maps in your atlas. How did you do?

18. Study the eastern "Lower 48 states" on the map and predict where orographic rainfall is the greatest in the <u>eastern</u> half of the country. Does the average annual precipitation map confirm your prediction?

Find Arlington, VA,, (near Washington, D.C.), and study the topography west of Arlington as far as the West Virginia-Ohio state line before leaving this map.

19. Examine the Raven map of West Virginia, keeping topography of adjacent states (especially Virginia and Maryland) in mind, to see if there are locations likely to be particularly wet or relatively dry because of orographic precipitation and rain shadows. Describe these locations using topographic features and place names.

Find these sites before you leave the West Virginia map: Charleston (38.4° N, 81.6° W), Pickens (38.7° N, 80.2° W), and Moorefield (39.1° N, 79.0° W)

Tables 6-1 and 6-2 show mean monthly and annual precipitation over the 1971–2000 climatic "normal" for five stations in the Virginias. The stations provide a west-to-east transect of the orographic effects on precipitation throughout the Central Appalachians. Data are from the NOAA National Climatic Data Center by way of the South Carolina Department of Natural Resources Southeast Regional Climate Center, Columbia, SC. Charleston data are from Yeager Regional Airport and Arlington data are from Reagan National Airport. Big Meadows is in Shenandoah National Park, near Luray, VA.

Table 6-1.   Monthly and annual precipitation along a Central Appalachian transect.

Location	Jan	Feb	Mar	Apr	May	Jun	Jul	Aug	Sep	Oct	Nov	Dec	Annual
Charleston WV	3.25	3.19	3.90	3.25	4.30	4.09	4.86	4.11	3.45	2.67	3.66	3.32	**44.05**
Pickens WV	5.74	5.03	6.10	5.43	6.24	5.78	6.73	5.29	4.95	4.25	5.23	5.49	**66.26**
Moorefield WV	1.99	1.75	2.57	2.32	3.59	3.46	3.63	3.47	2.92	2.74	2.51	1.93	**32.88**
Big Meadows VA	4.12	3.38	3.91	4.10	5.10	5.15	4.90	4.36	6.04	5.13	4.87	3.82	**54.88**
Arlington VA	3.21	2.63	3.60	2.77	3.82	3.13	3.66	3.44	3.79	3.22	3.03	3.05	**39.35**

Table 6-2.   Annual precipitation and snowfall data for five Mid-Atlantic sites.

Location	Annual Precipitation		Annual Snowfall	
	inches	cm	inches	cm
Charleston WV	**44.05**	**111.9**	*34.3*	
Pickens WV	**66.26**	**168.3**		*405.6*
Moorefield WV	**32.88**	**83.5**	*22.4*	*56.9*
Big Meadows VA	**54.88**	**139.4**	*44.0*	*111.8*
Arlington VA	**39.35**	**99.9**	*15.8*	*40.1*

20. Complete the snowfall columns by calculating the two missing metric or imperial equivalents in table 6-2. (Reminder: 1 inch = exactly 2.54 cm.)

21. Pickens averages 405.6 cm of snowfall a year, ranging from 250 to 675 cm from year to year. Hold your hand 405.6 cm above the floor.

22. OK, you can't reach that high; 405.6 cm would bury the top of the backboard on a regulation NCAA basketball court. How do you think Pickens deals with the issue of snow removal?

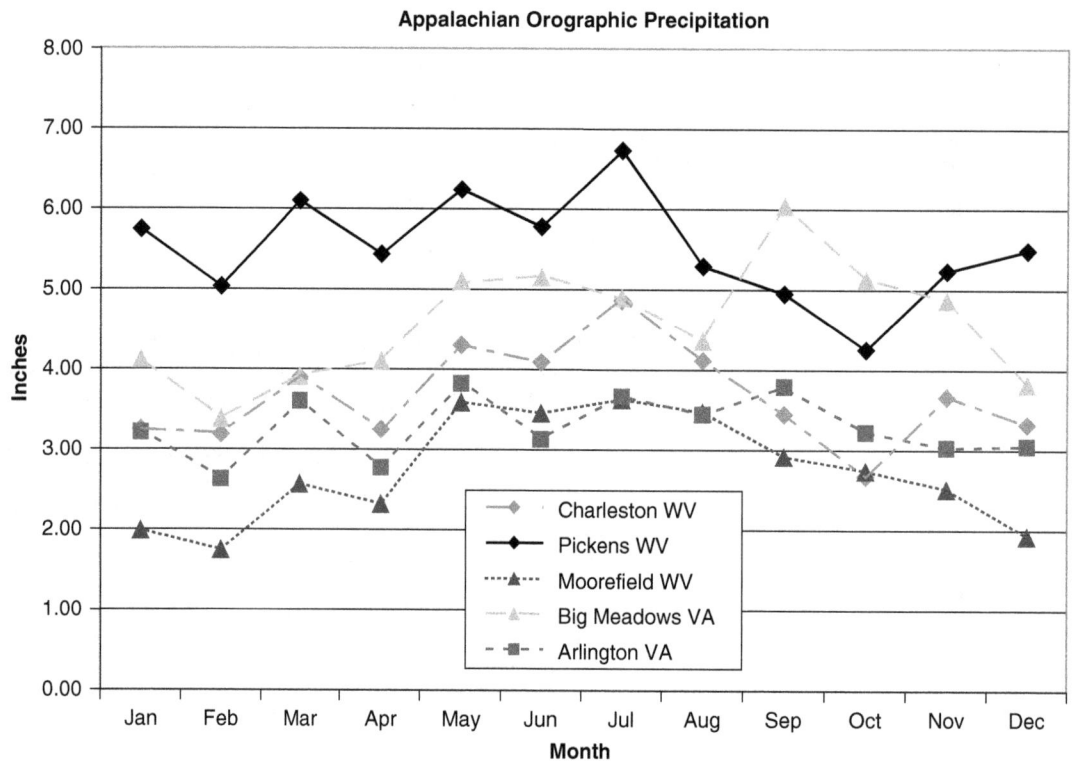

Figure 6-1 *Monthly precipitation plot across the Appalachians. Data are from table 6.1.*
Graph drawn by Steven Kite and Amy Hessl.

23. When are differences caused by orographic effects strongest in the Eastern U.S.? When are they weakest?

24. What months see normal precipitation at Big Meadows exceed that at Pickens?

25. What does this temporary ascendency by the Blue Ridge Mountain station suggest about the seasonal importance of Atlantic Ocean moisture vs. Gulf of Mexico moisture?

26. What meteorological phenomenon with great potential for heavy precipitation and very high winds appears with highest frequency at this time of year?

27. What type of weather dominates precipitation during the months when Pickens receives more than twice as much rain as Moorefield?

28. What type of weather dominates precipitation during the months when Moorefield receives an average of more than 3.00 inches of rain per month?

29. Label three periods on figure 6-1 indicating which style of precipitation dominates the trends: "Frontal," "Convectional" and "Hurricane."

Lab Exercise 7

Weather Forecasting and Climate Change

Readings

Strahler, **Introducing Physical Geography,** John Wiley & Sons, New York. **Chapter 6**.

Concepts

Before starting this lab make sure you understand these concepts:

- Air masses
- The geographic origins of air masses common in the Eastern U.S.
- Types of weather fronts: warm, cold, stationary, and occluded (= front aloft)
- Causes of precipitation
- Types of precipitation (frontal, convectional, hurricane/tropical cyclone)
- Statistical terms of average, mean, maximum, minimum, anomaly, and trend line

Materials

Access to a computer and the Internet, Strahler textbook, World Atlas.

Learning Objectives

Upon successful completion of this lab, students will be able to:

- Read and interpret weather maps
- Identify fronts, air masses
- Understand the difference between accuracy and precision
- Create graphs and tables using NOAA historical climate data
- Compare climate change at statewide, national, and global scales

On-Line Weather for Today

The next set of questions requires internet access, and include directions that may become outdated when websites are updated and redesigned.

Access the National Weather Service National Prediction Center web site: http://www.wpc.ncep.noaa.gov/index.php (viewed May 2019). **<u>Don't look at Day 2 (tomorrow) or Day 3 forecasts yet!</u>**

Examine the Overview map for Day 1. Note the location of surface high pressure (H) and surface low pressure (L). Remember winds in the Northern Hemisphere generally circulate clockwise around high pressure, and counter-clockwise around low pressure.

1. What fronts influence today's regional weather, if any?

2. Using terms from the assigned readings, explain what air mass or combination of air masses dominate today's local weather?

3. In what geographic region did this air mass or each of the air masses likely originate?

4. What fronts and air masses are in position to influence local weather in the next 24–48 hours? Where are these fronts right now?

5. Briefly describe regional conditions using the Overview and QPF (Quantitative Prediction Forecast) Day 1 maps to see where precipitation is likely to fall today.

Use NWS Graphical Forecast map https://graphical.weather.gov (viewed May 2019).

6. What measurement units and temperature scale are used on these maps?

Scroll down to the smaller interactive Additional Graphical Forecast Elements maps and click on the Temperature map. Note a larger map appears with an interactive table with that allows up to four time options for 14 different types data displayed in the map panel.

7. Toggle Today/Tonight (top row) and Max/Min Temperature (2nd row) to find

   - **today's forecast high temperature** _____.

   - **tonight's forecast low temperature** _____.

Slide back and forth through the 8 am, 11 am, 2 pm, and 5 pm options in Temperature, Wind Speed, and Sky Cover rows to see how these variables are expected to change through the day.

   Check the daily progression of Amount of Precipitation and Snow Amount for today.

8. Summarize your campus weather forecast for today in 280 characters or less.

9. Now that you are familiar with current conditions and the weather map, **form a hypothesis on what the local weather will be like for the next 48 hours, based on the current conditions only**. Use 280 or fewer characters.

Return to the NOAA-NWS National Forecast Chart www.wpc.ncep.noaa.gov/index.php—
Now jump ahead to Day 2 and Day 3 charts—thanks for your earlier patience!

10. How do the NOAA-NWS Day 2 and 3 National Forecast Charts differ from your hypothesized forecast? How were the two forecasts similar?

11. Explain what you might have overlooked or interpreted differently than the experts.

Go to the Weather Channel website at weather.com and enter the local zip code or town/city name. Click on "10 day" forecast.

12. Explain the difference between accuracy and precision.

13. How does the weather.com forecast compare to the 48 hour (Day 3) forecast made by NOAA-NWS.

Answer question 14 two days from now. Set your calendar reminder NOW.

14. **Finally, 48 hours later**, describe how your forecast held up compared to **what actually happened!** What might have caused the weather to pan out differently than you (or Weather.com) predicted?

Climate Change

More than 100 years of historical climate data exists for many stations in the United States. These data are available for free through NOAA National Centers for Environmental Information (NCEI) at https://www. ncdc.noaa.gov/data-access/land-based-station-data/land-based-datasets/global-historical-climatology-network-ghcn (viewed May 2019).

In this section, we will compare and analyze mean temperature and precipitation trends for several states in the United States, and assess what trends are shown by records from these states.

Open the NCEI "Climate at a Glance" web site https://www.ncdc.noaa.gov/cag/ (viewed May 2019) Click on the "Statewide" and "Time Series" tabs, then enter these search choices:

- Parameter: Average Temperature
- Time Scale: 12 Month
- Month: December
- Start Year: 1895
- End Year: the current year
- State: West Virginia
- Do <u>not</u> check "Display Base Period" or "Display Trend" (yet).

Click the blue "Plot" button. Give the system a minute or two to display a data table and create a plot of statewide annual mean temperature for more than a century. You may have to zoom out to see the entire plot.

15. Which is more useful for quick analysis: the table or the plot?

16. Describe any trends you see in annual mean temperatures for West Virginia?

The table output default is in chronological order. Click on the **"Rank" column header to re-sort the data from coldest to warmest.**

17. List the five coldest years recorded in NOAA data for West Virginia? How many of these are within your lifetime?

Click on the "Rank" column header again to re-sort the data from warmest to coldest.

18. List the five warmest years recorded in NOAA data for West Virginia? How many of these are within your lifetime?

19. Calculate the difference between the mean temperature in the warmest year and the mean temperature in the coldest year.

Return to the top of the page; keep all of the parameters the same, except **click on "Display Base Period"**, but do not check "Display Trend" (yet). Click on the blue "Plot" button, and patiently wait until a new plot appears, this time showing the long-term average temperature as a line across the total time period.

20. What is the statewide mean temperature in °F over the period since 1895?

21. Describe any trends in West Virginia temperatures you overlooked, but now see.

Return to the top of the page; keep all of the parameters the same, except **check "Display Base Period"**, **"Display Trend"**, and **"per Century"** before again clicking the "Plot" button and waiting. Note a **trend line** for the period appeared on the plot.

22. What is rate of mean annual temperature change since 1895?

Back to the top of the page; **change "Start Year", "Display Base Period-Start", and "Display Trend-Start" to 1975**. Keep other parameters the same… clicking the blue "Plot" button, and try to remember your parents' birth years to help the time pass.

23. What is the statewide mean temperature over the period since 1975?

24. What is rate of mean temperature change since 1975?

Return to the top of the page; keep all of the parameters the same, including the 1975 start dates, except **select Parameter = Precipitation**.

25. What is the statewide mean annual precipitation over the period since 1975?

26. What is rate of mean annual precipitation change since 1975?

Next, keep parameters the same, including precipitation, but **reset start dates to 1895**.

27. What is the statewide mean annual precipitation since 1895?

28. What is rate of mean annual precipitation change since 1895?

Click the "Rank" column header to sort the data from coldest to warmest, followed by re-sorting from warmest to coldest.

29. List the five wettest years recorded in NOAA data for West Virginia? How many of these are within your lifetime?

30. List the five driest years recorded by NOAA in West Virginia? How many of these are within your lifetime?

31. Which year really stands out as exceptionally dry?

32. After discussion with your classmates, describe the changes in West Virginia climate since 1895. Consider both annual temperature and precipitation and limit you well written answer to 280 characters or less.

Change the "State" parameter to a different state, possibly your home state or one in which a friend or relative lives. If you are a native of West Virginia, pick a state in a different region, and try not to pick the same state as others in your lab section.

Northern states like, Maine, Minnesota, and Alaska, have very striking temperature trends, but so do many others. In fact, most states have experienced more distinctive climate change trends than West Virginia and Kentucky.

33. Which state did you pick? _____

34. Write a short hypothesis on how climate change has differed in this state from what has been recorded in West Virginia.

Use the NOAA Climate at a Glance web page to run the annual mean temperature and annual precipitation plots for your selected state over the period since 1895.

35. What is the statewide mean temperature in °F over the period since 1895?

36. What is rate of mean annual temperature change since 1895?

37. What is the statewide mean annual precipitation over the period since 1895?

38. What is rate of mean annual precipitation change since 1895?

39. How do these data compare to West Virginia data?

40. Was the hypothesis you wrote in 34 tentatively confirmed or rejected.

41. Discuss the results from a variety of states to come to a group-wide consensus of whether or not climate change in the United States is significant, in 280 characters.

NOAA Climate at a Glance temperature and precipitation plots for the 48 contiguous United States on the following two pages may help your class reach a consensus.

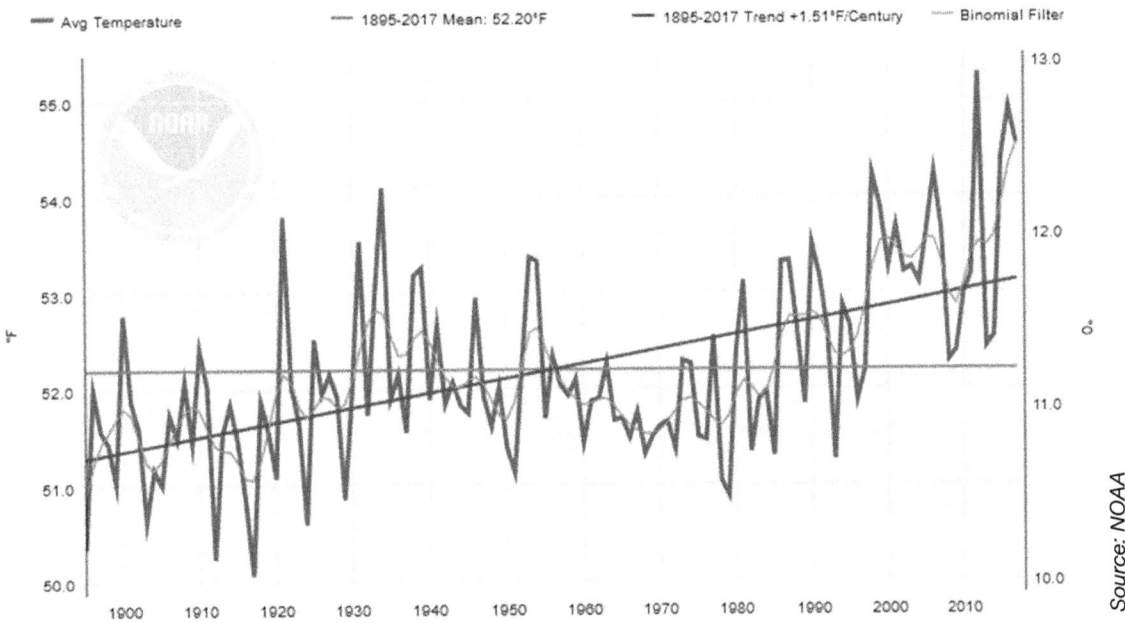

Figure 7-1 *Mean annual temperature, long-term average temperature, trend line, and binomial filtered trend for the 48 contiguous United States, 1895–2017.*

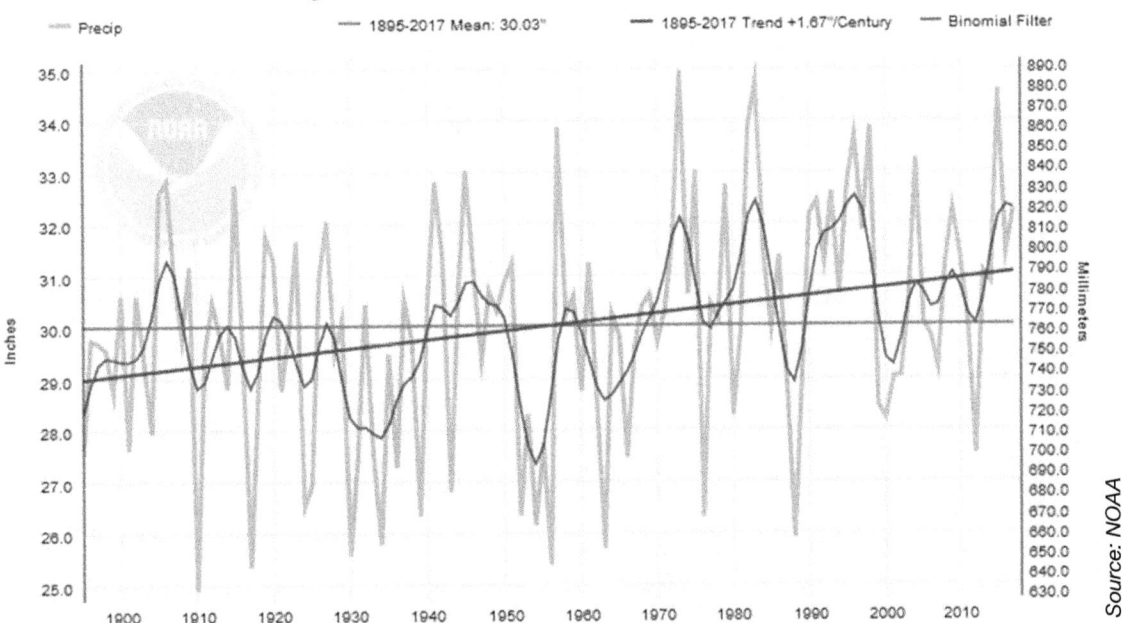

Figure 7-2 *Annual precipitation, long-term average precipitation, trend line, and binomial filtered trend for the 48 contiguous United States, 1895–2017.*

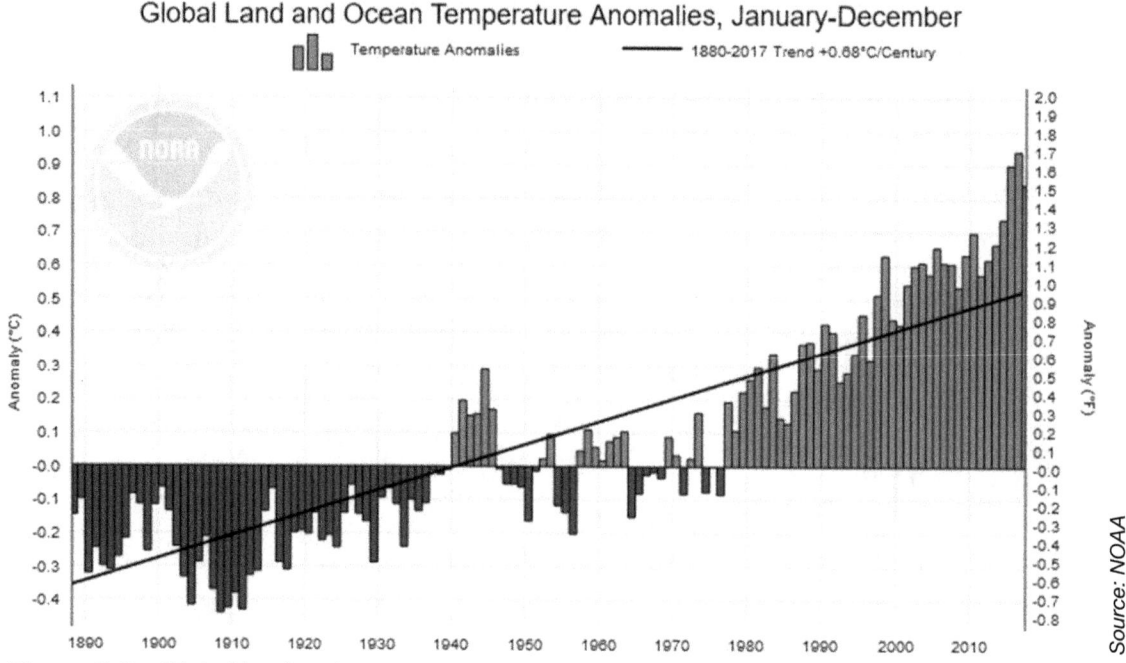

Figure 7-3   *Global land and ocean temperature anomalies 1880–2017.*

42. Was the group able to reach a consensus? What were the objections and sticking points for individuals who were unconvinced of what other class members saw in the data? How was the class able to overcome the obstacles, if they did, or why was the group unable to agree?

Lab Exercise 8

Analyzing Stream Flow Data with Microsoft Excel to Understand and Predict Floods

Reading and Tutorial assignments to be completed <u>BEFORE</u> Lab 8

Readings

Strahler, A.H., 2013, **Introducing Physical Geography**, John Wiley & Sons, New York, **Chapter 14**, *or alternative reading assigned in GEOG 107.*
How Does a U.S. Geological Survey Streamgage Work? (2 page fact sheet pdf file) http://pubs.usgs.gov/fs/2011/3001/pdf/fs2011-3001.pdf (viewed May 2019).

Materials Used in Lab

- Access to a computer running Microsoft Excel and Internet access.

Learning Objectives

Upon successful completion of this lab, students will be able to:

- Use Microsoft Excel software to manipulate and analyze stream-gage data
- Retrieve and manipulate USGS stream-flow records
- Calculate the probability of exceedance and recurrence interval for floods of various magnitudes
- Develop hypotheses about flood likelihood based on historical stream flow data
- Examine the concept of a "100-year flood" and discuss the problems and limitations of this term
- Discuss human interaction with streams and flooding

Before starting the lab, **be sure you understand** these terms: **stream, watershed, hydrograph, peak flow, rating curve, flood, 100-year flood,** and **floodplain.** Look up unfamiliar terms in the USGS <u>Water Science Glossary of Terms</u> site at http://water.usgs.gov/edu/dictionary.html or the USGS <u>Floods: Recurrence intervals and 100-year floods</u> site at http://water.usgs.gov/edu/100yearflood.html. (Both viewed May 2019)

Microsoft Excel Tutorials

Microsoft Excel is software designed to create **spreadsheets**, data tables used to record, explore, and analyze information. A spreadsheet includes a grid of **cells** arranged in numbered rows and letter-named columns. Each cell contains a piece of information (such as a stream flow measurement), while the row number and column letter for each cell provide a unique address for the information that can be linked link to other information (such as the time and location for that flow reading).

You will use Excel in lab exercises to sort data, create new types of data using math tools called **macros**, and create charts or graphs. Compared to boring columns and rows of numbers, charts and graphs allow most people more clearly see and understand trends in geographic data.

Excel has many important applications outside of physical geography, such as budgets and finance, so this software is a worthwhile real-world tool to learn and master. Although many GEOG 106 students may be expert at Excel, it will be new to others. Those with Excel experience may want to review the software, or just jump right in!

If you have not used Excel or your knowledge of the software is rusty, a review of on-line software tutorials should help you complete labs in a timely manner and get more out of the exercise. Take tutorials if you are unfamiliar how to **Sort** rows and make charts with Excel, or unsure how to use these macros: **=Average, =Count, =Max, =Min, =Sum**.

Here are some tutorial resources viewed when the manual was revised in May 2019:

What is Excel (links to useful sites based on older Excel versions)—
https://serc.carleton.edu/introgeo/mathstatmodels/UsingXL.html

How to Use Excel (links to sites based on older Excel versions, including a "cheat sheet")
https://serc.carleton.edu/introgeo/mathstatmodels/xlhowto.html

Basics tasks in Excel—https://support.office.com/en-us/article/basic-tasks-in-excel-dc775dd1-fa52-430f-9c3c-d998d1735fca

The Beginner's Guide to Excel—Excel Basics 2017 Tutorial (22 minute video) https://www.youtube.com/watch?v=rwbho0CgEAE

Excel Tutorial: Learn Excel in 30 Minutes—Just Right for your New Job Application (30 minute video)
https://www.youtube.com/watch?v=7RCdzTpKO0A

Beware web links may "rot" in time, so to see current resources you may need to look at updated GEOG 106 web resources or use key words searches in a web search engine.

Stream-Flow Data

The United States Geological Survey (USGS) maintains a system of stream gages that provides current and long-term data on stream flow throughout the United States. Current data typically are recorded and stored at gage stations every 15 to 60 minutes, then promptly relayed via satellite, telephone lines, or radio to the USGS. The USGS strives to make stream flow data available for viewing within minutes of arrival.

Several technical terms are crucial in understanding floods:

<u>Recurrence Interval</u>—The predicted average amount of time between flows of a specified discharge (or stage).

<u>Annual Exceedance Probability</u>—Chance that a flow of a specified discharge (or stage) or larger will occur in a given year.

<u>Discharge</u>—Volume of flow per time, given in cubic meters per second (over most of world), cubic feet per second, or odd units like acre-feet per day (in the U.S.).

<u>Stage</u>—Height of water surface above an arbitrary reference level at a stream gage.

Estimation of recurrence interval and annual exceedance probability has been practiced for decades, but historic flood records may be poor predictors of future floods in watersheds that have seen significant recent changes in character, like damming, deforestation, urbanization, or exceptional climate change.

Downloading Stream-Flow Data

Use an internet browser to go to USGS Current Water Data for the Nation—Daily Streamflow Conditions at http://waterdata.usgs.gov/nwis/rt.

Put your curser over West Virginia on the map to bring up a detailed map of gages within the state. Click on the Morgantown area to see where nearby current water data stream gages are located.

1. What gage station is the closest to the WVU Downtown Campus? *(Hint: it is NOT on the Monongahela River, but is on a tributary creek).*

2. Click on this station and scroll down to the two hydrographs for this gage; what are the current stream discharge and stage?

3. What were the highest discharge and stage recorded during the last seven days?

Go to the upper right of the page to change "Days" setting from the default of 7 to 90.

4. What were the highest discharge and stage recorded over the last three months?

A Tragic Flood

Talk about floods with older folks in eastern and north-central West Virginia and it will not take long for them to mention the Flood of 1985. In fact, there was a long-standing local Morgantown rock band called '85 *Flood*. Property damage in West Virginia exceeded $600,000,000 (in 1985 dollars) in the worst disaster in the USA for the year. Large portions of the towns of Rowlesburg, Parsons, Franklin, Petersburg, and Moorefield were devastated when the Cheat and South Branch Potomac rivers crested 5 to 10 feet higher than any flood in memory, Thousands of homes and businesses were destroyed or damaged; almost none were covered by flood insurance.

Many other communities in the Mid-Atlantic region were severely impacted; some never to fully recover. Search "Potomac Flood 1985" on YouTube to view old TV newscasts, including the ABC News story. The 1985 flood was tragic, but was it the greatest flood on record everywhere in the state? Was it a unique "Act of God" or a predictable event?

Exploring Peak Discharge Data from the Potomac River Using Microsoft Excel

The magnitude of a flood is usually measured by its **peak flow**. **Annual peak flow** is the highest discharge flowing at a site during a year. As you might expect, large floods occur less frequently than small ones. Summaries of peak flow data help hydrologists and the public fathom the likelihood and extent of flooding.

Several calculations help revel the history of flooding and assist in forecasting future floods. These include **historic maximum peak flow,** and the **recurrence interval** and **probability** of a flood of a certain magnitude in a given time period. If one knows the discharge at which flood damage begins, one can predict the average time interval between damaging floods, a key issue in planning flood recovery.

The next section of the exercise focuses on high flows at one of the longer stream gage records in the region: the USGS gage on the Potomac River at Shepherdstown, West Virginia, about 90 km (56 miles) northwest of the Capital in Washington, D.C. The USGS recorded flows continuously at this site from 1929 until 1993. Other flow data are available from this site before and after this period of continuous recording, but the exercise will stick to 1929–1993 to keep things simple.

Flows on many rivers in the conterminous US are typically at their lowest in early autumn, so the USGS determines annual peak flows, not by calendar year, but by a **water year** that runs from October 1 until September 30.

Note that many smaller annual peak flows in the USGS data do not rise above of the riverbanks, and are not true floods in the usual sense of the word.

Open the "Potomac_Sheperdstown_Peak.xlsx" peak flow data file using Microsoft Excel by double clicking on the file icon on the lab computer desktop. You may have to click on the "1929–1993 Peak Flow Data" tab first, but when it opens the file looks like this:

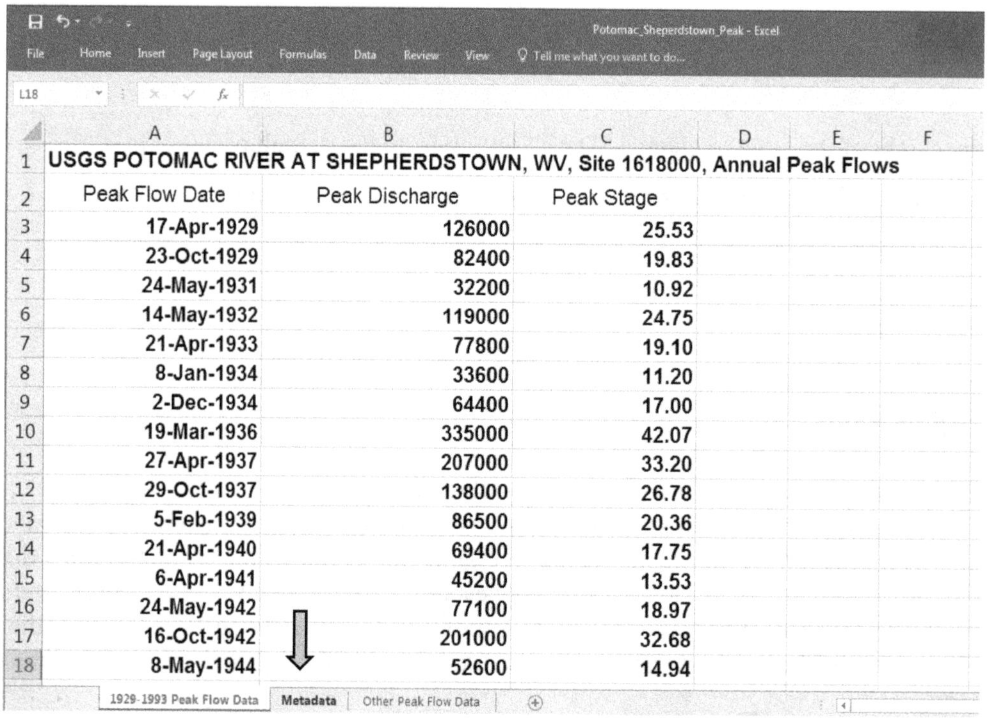

	A	B	C
1	USGS POTOMAC RIVER AT SHEPHERDSTOWN, WV, Site 1618000, Annual Peak Flows		
2	Peak Flow Date	Peak Discharge	Peak Stage
3	17-Apr-1929	126000	25.53
4	23-Oct-1929	82400	19.83
5	24-May-1931	32200	10.92
6	14-May-1932	119000	24.75
7	21-Apr-1933	77800	19.10
8	8-Jan-1934	33600	11.20
9	2-Dec-1934	64400	17.00
10	19-Mar-1936	335000	42.07
11	27-Apr-1937	207000	33.20
12	29-Oct-1937	138000	26.78
13	5-Feb-1939	86500	20.36
14	21-Apr-1940	69400	17.75
15	6-Apr-1941	45200	13.53
16	24-May-1942	77100	18.97
17	16-Oct-1942	201000	32.68
18	8-May-1944	52600	14.94

Figure 8-1 *Excel file of Potomac River annual peaks flows for 1929–1993. Arrow points to the metadata tab.*

5. Click the metadata tab at the bottom to find out what each column stands for. (Many fields have been deleted to simplify your dataset, bolded items have been included.) Define these three column headings, including measurement units, in the space below each heading.

Peak flow date # peak_dt	Peak discharge # peak_va	Peak stage # gage_ht

6. Remembering the concept of <u>water year</u>, explain why some <u>calendar years</u> either have no events (e.g. 1930 & 1943) or contain two event entries (e.g. 1929 & 1942),

Generate Hypotheses About Flood Frequency

7. What is a hypothesis?

8. Remembering that the 1985 flood has been called the worst natural disaster in West Virginia history, formulate and write a simple hypothesis regarding the magnitude and importance of the 1985 flood in the Shepherdstown area.

Identify the Maximum, Minimum, and Mean Values of Peak Flow

Using the scroll bar, scroll down to the bottom of the data set and click on the cell below the last record for Peak Discharge. Then, with that cell highlighted, go up to the summation symbol (Σ) on the Formula tab (top of screen) and click on the down arrow (triangle) to give you more options. These **macro** options include COUNT, SUM, AVERAGE, MINIMUM, and MAXIMUM. Click first on AVERAGE, and then hit return.

9. Using the same technique, find the count, sum, minimum, and maximum of values in this column. Record the values calculated using macros in the table below:

Average	Count	Sum	Min	Max

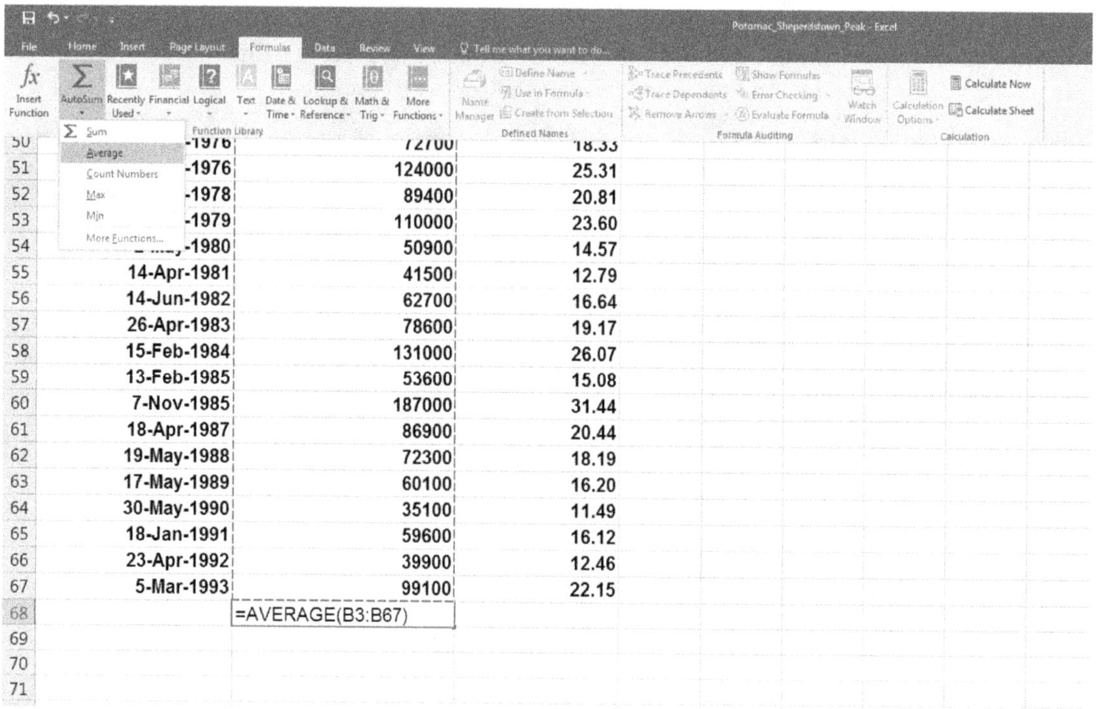

Figure 8-2   *Calculating a formula for average in Excel.*

10. What might you miss by just looking at the average? How might climate or land-use change over time affect peak flow?

Graph Peak Flow Through the Years

- Select all the data in columns A and B (leaving off the title of the data file, but not the column headings).

- With data in columns A and B selected, click the **Insert** tab and click the **Chart** icon. Then select a "**Scatter (X, Y)**" chart, opting for a simple plot without lines.

Enlarge the chart for better viewing by clicking and dragging a corner.

11. Has annual peak flow changed over time? If so, how?

12. Would your answer to the last question change if the March 1936 flood were not part of the record?

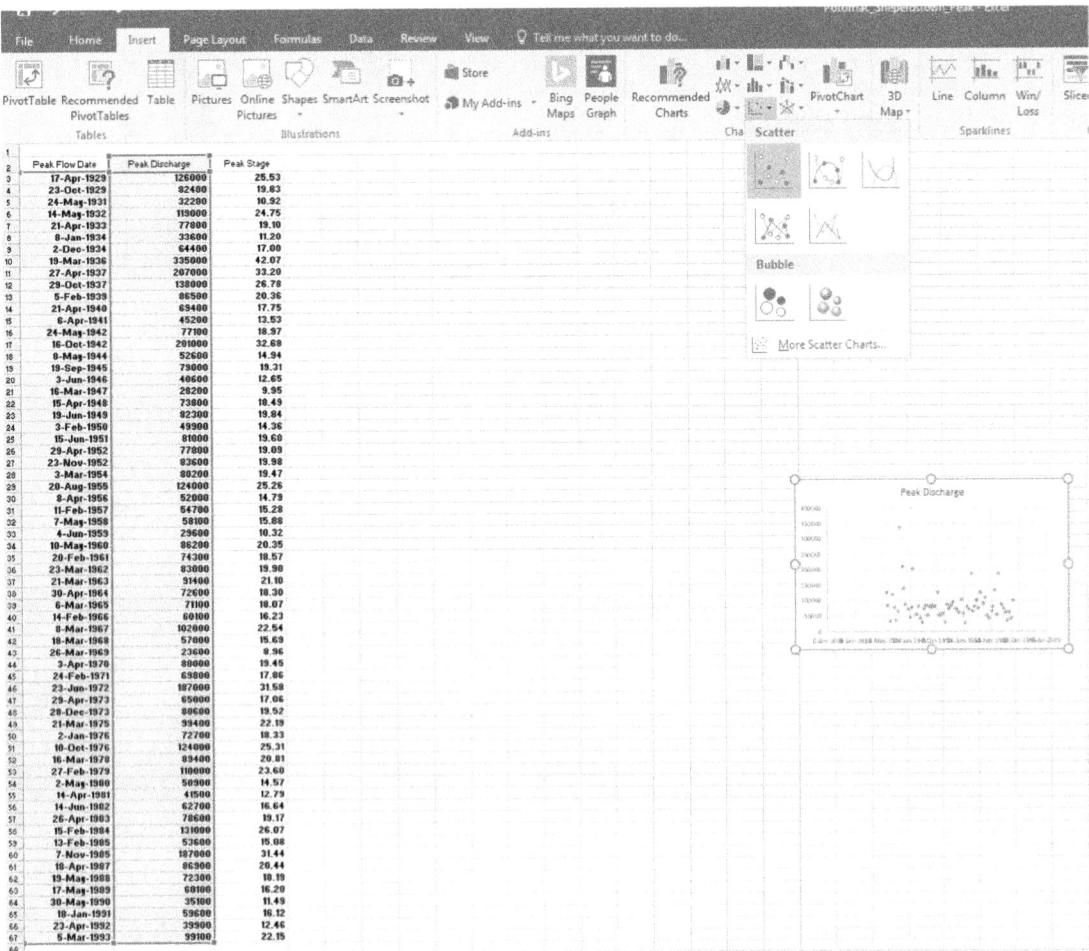

Figure 8-3   *Creating a scatter chart in Excel.*

Flood Frequency Analysis

Graphs can be useful for understanding quantitative data, but alone they are not ideal tools for flood forecasting. For example, by looking at the graph, it may be difficult to determine how often a flood of a particular magnitude has occurred during the 65-year record or if and how the frequency of a certain size flood has changed over time.

Determining flood recurrence interval for past events is straightforward. The recurrence interval (**T**) is associated with rank and length of record by the equation

$$T = (n+1)/m$$

where **n** is the number of years of record and **m** is the magnitude ranking of each event.

Potomac River data can be sorted and ranked by Peak Discharge, with highest flow at top; then, each flow's recurrence interval can be calculated, using n = 65, because we have 65 years of data. As an example, here are the ranks and recurrence intervals for the first five observations in the dataset (based on statistics for the entire dataset).

Peak Flow Date	Peak Discharge	Rank	T
4/17/1929	126000	7	9.43
10/23/1929	82400	20	3.30
5/24/1931	32200	59	1.12
5/14/1932	119000	9	7.33
4/21/1933	77800	28	2.36

Flood Probability

Like gambling on sports or cards, to make general flood forecasts, we assume the past tells us something about what the future, and then apply probability statistics.

The probability (**P**) of an event with recurrence interval **T** is

$$P = 1/T$$

The probability P_T that a given event will be equaled or exceeded at least once in the next **r**-years is:

$$P_T = 1 - (1 - P)^n$$

where n = number of years.

So for the first few observations, the probabilities of each of these events will be equaled or exceeded in the next 100, 25, and 10 years is:

Cal. Year	Peak Flow Date	Peak Discharge	Rank	T	P	Pt (100)	Pt (25)	Pt (10)
1929	10/23/1929	82400	20	3.30	0.30	100 %	100%	97%
1929	4/17/1929	126000	7	9.43	0.11	100 %	94%	67%
1931	5/24/1931	32200	59	1.12	0.89	100 %	100%	100%
1932	5/14/1932	119000	9	7.33	0.14	100 %	97%	77%
1933	4/21/1933	77800	28	2.36	0.42	100 %	100%	100%

A probability of 1.00 = a 100% chance. A probability of 0.97 = a 97% chance

Now that you have seen what can be done for a small data set, it is your turn to use Excel to rank annual flow, determine its recurrence interval, and the probability of a flood will equal or exceed this one in the next 1, 10, 25, and 100 years.

13. Calculate the Rank, T, P and exceedance probability for the entire dataset, including probabilities that each event will be equaled or exceeded in 100, 50, and 10 years. *Your instructor may ask you to submit your table as an Excel worksheet attachment.*

How to do the Calculations in Excel

Step 1. Sort the peak flow data by discharge.

- Select all of the rows that include flow data.

- Click on the Data near the top of the screen, and then click on the sort icon.

- Select sort by "Peak Discharge" (Column B) in "Largest to Smallest" order.

Step 2. ***Create a Rank Column***

- Enter the word "Rank" in the cell right of "Peak Stage".
- Enter a value of 1 in the first cell below the word "Rank".
- Highlight a column from this 1 value down to the last row of data.
- Enter the words "**Fill Series**" after the "Tell me what you want to do . . . " lightbulb.
- When prompted select "Series", Check "Columns", "Linear", and Step value = 1.

Step 3. ***Create a T (Recurrence Interval) Column***

- Enter the letter "T" in the cell right of "Rank".
- Assuming rank number 1 is in cell D3, click on cell E3 and enter the following macro formula in the formula bar above the spreadsheet:

$$= ((65+1)/D3)$$

- Cell E3 should read "66", but the formula bar will show the macro equation.
- **Select Cell E3 and all of the cells down to the smallest flow value;** then use "**Fill Down**" (after the lightbulb) to calculate T values for all of the other floods. Although the macro is the same over the whole column, *the calculated numbers should have different values if you fill the column correctly.*
- Clean up the column display and inconsistent number of digits by selecting the top of the E column, clicking on the Home tab, and selecting "Number".

Step 4: ***Create a P (Probability) Column***

- Enter the letter "P" in the first cell to the right of the letter "T".
- Assuming a calculated value for "T" is in cell E3, click cell F3 and enter this macro in the formula bar:

$$=(1/E3)$$

- Select Cell F3 and all of the cells down to the cell beside the smallest T value; then use "Fill Down" (after the lightbulb) to calculate P values for all of the other floods.
- Clean up the F column using "Home" and selecting "Percentage".

Steps 5, 6 & 7. ***Create Pt (100), Pt (25), and Pt (10) Columns***

- Following previous column additions, create column headings for probability of recurring in these three time intervals, and enter the following macros.

$$=(1-(1-F3)^\wedge 100)\text{ in cell G3}$$
$$=(1-(1-F3)^\wedge 25)\text{ in cell H3}$$
$$=(1-(1-F3)^\wedge 10)\text{ in cell I3}$$

- Use "Fill Down" and "Percentage" as before.

Questions About the Results of You Anlaysis

14. How did 1985 rank during 1929–1993 on Potomac River at Shepherdstown?

15. What years had higher peak flows (if any)?

16. How likely is it that a flood with the magnitude of the 1985 flood at this gage will recur or be exceeded in the next 10 years? _____ Next 25 years? _____ Next 100 years? _____

17. Are peak flows of the 1980's unusual?

18. What multiple-year time period in the record is most atypical and least likely to be repeated?

19. Was your hypothesis regarding the magnitude of the 1985 flood in the Shepherdstown area correct?

20. Explain why or why not.

Falling Run Field Trip
By J. Steven Kite & Mitzy L. Schaney

Assignment to be completed <u>BEFORE</u> Lab 9

Readings

Browse the 50+ photos that appear in response to the search term "Falling Run" at the West Virginia University Libraries History on View web site http://wvhistoryonview.org (retrieved May 2019) before the trip. Dating between the 1880s and 1970s, most of the photos show Falling Run and its watershed, which lies wholly within 2.2 km of Brooks Hall. Images include a small waterfall, a scary pedestrian bridge, the construction of old Mountaineer Field, the forerunners to modern streets, and many still-standing or long-gone private houses and WVU buildings (including Hick House). Note the increase in real-estate development and other land-use changes through time.

Materials for the field trip

Wear long pants and closed-toe shoes that may get wet and dirty.

Learning Objectives

Upon successful completion of this lab, the student should be able to:

- Identify and discuss environmental issues related to urban streams.
- Identify indicators of slope instability and landslide mitigation.
- Formulate hypothesis over the future of the Falling Run valley.

Falling Run—A Stream Lurking Under the WVU Downtown Campus

Falling Run and its valley (figure 9-1) was a pleasant enclave on the edge of campus in the early decades of WVU. The History on View photos show a 3 m high waterfall that gave the stream its name was very popular for special occasion photos. Regrettably, green space did not have high priority in 1924, when the lower reach of Falling Run was routed into a concrete box culvert under Mountaineer Field. The hidden stream lies out of site and out of mind to most of us, although it still flows within 30 m of Brooks Hall.

While lower Falling Run is ecologically alarming, upper parts of the watershed in the WVU Organic Farm remain in fair condition. Much of the rugged middle portion of the valley escaped the haphazard development that swept Morgantown in the 20th Century, and was purchased by WVU in 2012. Since then, student volunteers have shepherded a trail system and relatively intact green space. Photos suggest there are more trees in Falling Run valley now than any time since the late 19th Century.

Book learning and lab work are very important, but it is impossible to truly understand landscapes and ecosystems without going out to see them in person. Many of you already spend a lot of time out in nature, while some may be surprised to see how much there is to see when you go outdoors and take the time to look.

We very strongly recommend **long pants and closed-toe shoes** for this 2 km (just over a mile) round-trip walking class exercise. **Beware of traffic everywhere and poison ivy in wooded areas. Avoid contact with stream water**; some orphan sewer lines in the watershed appear to discharge into the stream.

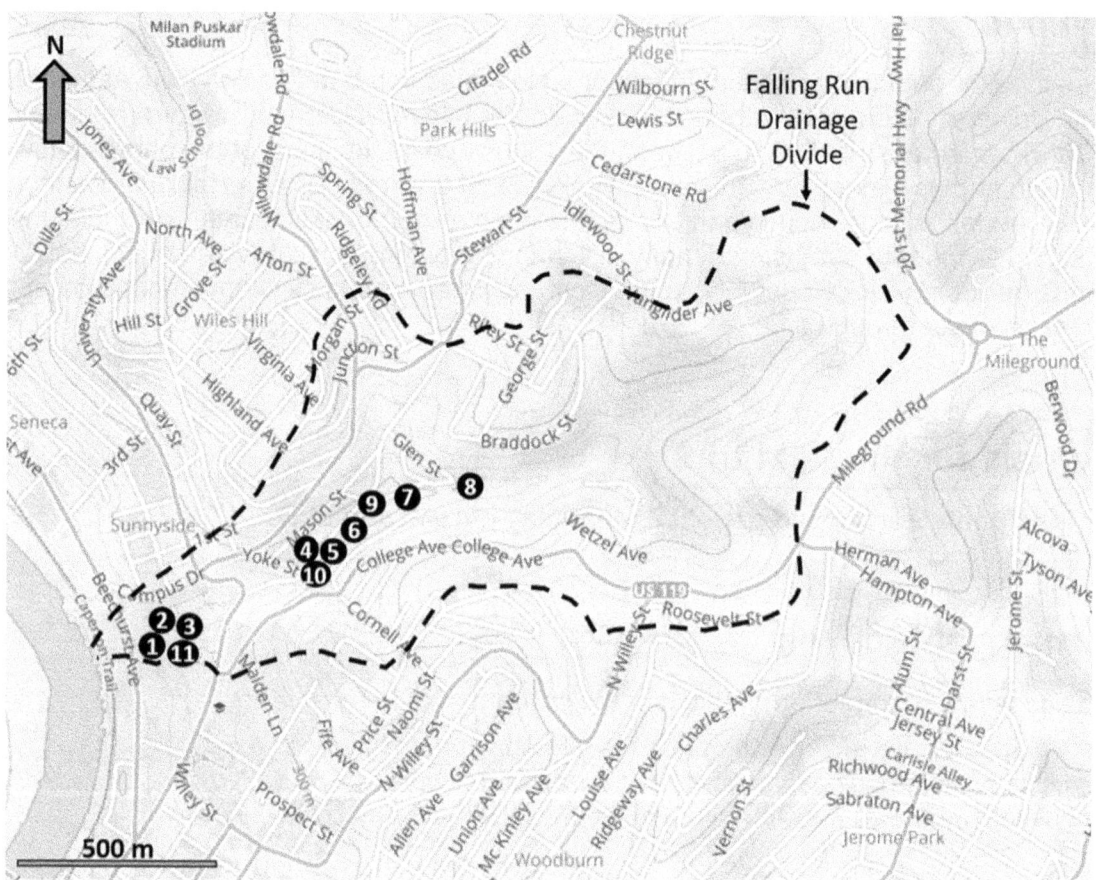

Figure 9-1   *Falling Run watershed, showing its drainage divide and our field trip stops. Stop 12 is not shown, but is virtually the same location as Stop 1. The drainage divide is the boundary separating land surfaces that drain into Falling Run from those that flow into some other stream. Base map was created using the USGS topoView web site at https://ngmdb.usgs.gov/topoview/ (retrieved May 2019).*

The trip begins in the usual classroom, treks along lower of Falling Run, and returns to a last stop on the Brooks Hall 5th floor. Stop locations appear in figure 9-1. The itinerary may vary depending on weather and other factors. **Follow directions. Do not leave valuables in the classroom. Do not leave the group without telling the trip leader**.

Stop 1.   Brooks Hall north entrance—The multi-colored landscaping gravel next to the entrance was mined from bed of the Ohio River, where sediment was transported by Ice Age glaciers and meltwater from bedrock tens or hundreds of kilometers away. The well-rounded shapes of the cobbles and pebbles confirm they have traveled far because stream transportation both reduces particle size and rounds originally angular rock fragments. Sedimentary rocks that make up most of the gravel likely came from bedrock within the upper Ohio River basin. Igneous and metamorphic rocks in the gravel originated from bedrock in Canada and carried into the Ohio River basin by glaciers.

Larger light-gray to brownish-gray cobbles and small boulders in the foyer are of local origin. A 2017 steam-line excavation in Woodburn Circle (near Stop 11) unearthed these large particles from a deposit that Geology & Geography Department founder I.C. White studied in the 1880s. *(Mountaineers Go First—Dr. White was a preparatory class student when WVU opened its doors in 1867; he grew to be one of the most influential geologists in North America during the late 1800s and early 1900s.)*

The sandstone cobbles and boulders originated from Morgantown area bedrock, but their large size and typical roundness show transportation a kilometer or more by a large stream about the size of the modern Monongahela River. They were found at an altitude ~ 50 meters (~165 feet) above the modern riverbed, indicating they are remnants of an ancient **river terrace** left high and dry when the river eroded down to its modern level. The particles have weathering rinds and iron stains suggesting they have weathered for at least 130,000 years, and may have been deposited on the river terrace as long as 1,000,000 years ago.

- To put the timing of the local cobbles and boulders in perspective, calculate how much time elapsed between creation of the terrace deposits and the invention of writing around 5100 years ago.

- Bedrock under downtown Morgantown is about 305,000,000 years old. How old was the bedrock when the cobbles and boulders came to rest on the old terrace?

Stop 2.   Falling Run undercover—This dry stream valley is an example of urbanization of landscape and the loss of green space. Falling Run once was a naturally flowing stream that emptied into the Mon River downhill from where we stand. When the old Mountaineer Field was built in 1924, the lower third of Falling Run, including its pretty waterfall, was put in an underground concrete culvert. The stream continued to be urbanized, and its lower reach is now completely underground. The best visible evidence of lower Falling Run is an alignment of manhole covers and drain grates approximating were it lies beneath the surface.

- What are some special hazards on urbanized streams?

- Would a "natural" stream running through campus improve WVU's aesthetic appeal?

Stop 3.   Hick House—This a spooky story verbatim from the WVU School of Medicine at https://medicine.hsc.wvu.edu/about/school-history/ (retrieved June 2018).

The HICK HOUSE was the first medical school building in the State of West Virginia. Constructed in 1892 in Falling Run Hollow below Woodburn Hall, it was intended to provide space for dissection of about 10 cadavers. Dr. James Hartington, appointed to the university faculty in 1887 as the first full time Professor of Anatomy, Physiology and Hygiene secured the $350 needed to construct the 14" × 20" building. It served Medical instruction for about ten years and provided material for area storytellers for many more. The origins *of the name "Hick" are not known, but it probably represents a local corruption of the Latin burial phrase "Hic Jacet" (here lies...). Whether true or not, cadavers here were known as "hicks," rather than "stiffs."*

- Formulate a hypothesis explaining why cadavers were stored and dissected in Hick House rather than in one of the other academic buildings on campus.

Continue eastward up Falling Run valley. Be alert and very careful crossing University Avenue at the crosswalk. Stay to the left side of Falling Run Road to Vandalia Hall.

Stop 4. **Storm-water detention basin**—The elongate depression between Parking Area 5 and Falling Run Road is usually dry, but may fill with water after rainfall. Detention basins only hold water for a short period before slowly releasing it into a stream or groundwater, while **retention ponds** generally pool water all the time. These two basins types can help manage runoff and trap sediment from urbanized areas, thereby preventing flooding, erosion, and deposition downstream.

- Examine the detention basin to explain how it is designed to fill with runoff and how it is supposed to drain slowly after filling.

Carefully cross to the right side of Falling Run Road at the east end of Parking Area 65 to walk on a wider, safer sidewalk.

Stop 5. Culverts—Observe the box and circular culverts behind and on either side of a house adjacent to Falling Run. Note their relative opening size. (The culverts may be difficult for the whole class to see at the same time, especially if vegetation is dense at the time of this trip.)

- Where will stream flow be most restricted during high runoff?

- What will happen if the flow is larger than the smallest culvert can handle?

Stop 6. A missing stream and **creep trees** at Parking Area 3.

- Would one be comfortable parking in this lot when the weather forecast includes a flashflood warning? Explain.

Note that trees on the hillside have bowed trunks deformed by **creep**, a very slow, silly-putty-like downslope progression of soil or weak rock. Tree trunks tilt when the soils in which they are rooted move downhill, so to avoid falling over the trees must grow asymmetrically to keep most of their trunks upright.

Creep and landslides are more likely to occur in soils or bedrock containing **shrink-swell clays**, which expand when wet, then contract upon drying. Shrink-swell clays are common in the soils and bedrock in Western Pennsylvania and North-Central West Virginia, including the Morgantown area.

The downslope rate of soil creep depends upon four factors:

- steepness (gradient) of the slope

- water absorption and content

- type of sediment and material

- vegetation

These factors are similar to those listed for landslides in your GEOG 107 textbook.

- Predict what might happen if the vegetation on the hillside were removed without any offsetting slope mitigation or protection?

Stop 7. A single-species stand of an **exotic invasive plant**—Japanese knotweed, *Polygonum cuspidatum*, is a troublesome aggressive plant. Introduced from Asia as an ornamental in the 1800s, it has a tremendous capacity for prolific reproduction and spreading along disturbed area such as floodplains and rail trails, out-competing and crowding out North American native plants. Japanese knotweed stands have very low diversity, negatively affecting almost every plant or animal species that would normally occur.

- Why is it detrimental to have very low plant and animal diversity?

- Once established, *Polygonum cuspidatum* is extremely difficult to eradicate. Formulate a strategy to keep this invasive species from taking over a green space.

Continue walking up Falling Run Road to an unmarked trail at the end of the road. Continue into the forest to see a relatively intact stream reach.

Stop 8. Falling Run revealed—**<u>Stay out of the water</u>**; although you may not see acid mine drainage (the source of the orange-brown color of many Appalachian streams), there are upstream "straight pipes" (direct sewage outflows).

- Are the cobbles in this streambed well rounded? Why or why not?

- How might unpermitted sewage discharge influence the stream ecology?

- List any other problems you or classmates recognize on this reach of Falling Run?

We have little time today to study the green space, so we encourage you to come back again and explore. Turn around. Head back down the valley, always alert to traffic.

Stop 9.   Retaining walls—Along our route back, observe a variety of good and bad retaining walls on the north side of Falling Run Road and Protzman Street.

- What happened to the excavated slope that was unprotected by a retaining wall?

- What attributes make a good retaining wall?

Stop 10.   Limestone **riprap**—The coarse angular rocks were placed here to protect a *ca.* 2008 landslide from further failure or gully erosion. This Greenbrier Formation limestone comes from one of several quarries 10 or more km east of Morgantown.

- Does the landslide appear stable?

- Can you see any other nearby landslides or washouts on this hillslope?

- Describe any enduring evidence of a small February 2018 debris flow that occurred about 100 meters west of the limestone riprap, near lamppost 357 next to the asphalt sidewalk leading to Ming Hsieh Hall.

- What can be done to diminish the long-term slope failure risk in Falling Run valley?

Continue up the sidewalk to the Ming Hsieh Hall deck, and then use the pedestrian bridge to cross University Avenue. Walk between Chitwood Hall and the College of Business and Economics Building.

Pause briefly by the ramp on the left that leads to the front of Woodburn Hall. The steam line excavation that uncovered old terrace boulders and cobbles discussed at **Stop 1** ran along the path of the rebuilt sidewalk in front of Woodburn. If we have time, we may look for newer sections of concrete that mark the path of digging in 2017.

Proceed to the right of Woodburn to the horseshoe-shaped overlook behind the hall.

Stop 11. Monongahela River valley overlook—Monongahela River is controlled by a lock and dam system to ensure navigation depth for commercial barges. The river flows north to Pittsburgh where it meets the slightly larger Allegheny River, which drains northern Pennsylvania and southwestern New York. Together the Monongahela and Allegheny form the Ohio River; hence, our campus is in the Ohio River Watershed.

- What is the English translation of the Algonquian name "Monongahela"?

The landslide visible beneath the Engineering Building is actually adjacent to the WVU Evansdale Campus greenhouse. The apparent proximity to the Engineering Building is an optical illusion because that is a 12-story building.

Some slope stability facts to ponder concerning this landslide:

- The road below the landslide, Monongahela Boulevard, was built by cut and fill along the base of this slope in the 1950s.

- Shale and fine grained rocks on this slope have been eroded progressively by freeze-thaw, rock fall, gullying, slips, debris flows, and tree tip; at times accumulating in deposits that must be mechanically removed to keep the road from being blocked.

- This progressive erosion of shale and fine grained rocks has over-steepened the slopes and left the resistant sandstones "cantilevered" in an unstable position.

- In January 1983, a stability threshold was exceeded and the slope failed catastrophically, blocking all four lanes of the boulevard and the PRT track.

- Rock fall has been very common since the big 1983 landslide. A large boulder that fell on 10 March 1994 weighed over 60 tons and appeared on page 1 of USA Today. Other large rock falls occurred in April 1994 and spring 1996.

- In 1997, a bulldozer was used to create a bench about 1/3 the way up the slope, in order to catch falling boulders.

- Unfortunately, the bulldozing activity further over-steepened the slopes and another big block came down in 2012.

- Several significant debris flows have occurred since 2012, including a lane-blocking event in February 2018.

- There are many landslide problems in the Morgantown area; the worst of all may be on River Road in Westover.

- How could the Monongahela Boulevard landslide problem have been avoided?

- How would you propose to address this slope stability problem, remembering that money to fix things does not grow on trees?

Stop 12. Brooks Hall **green roof**—Added in a major 2007 building remodeling, this was the first green roof constructed on the WVU campus.

A green roof (a.k.a. eco-roof or living roof) consists of live vegetation, planted in a soil-like growth medium, above a waterproof membrane. A green roof may also include other layers, such as a root barrier, a drainage system, or irrigation fixtures. The six species of low maintenance sedum and other plants on this green roof are drought tolerant; water on a roof can be scarce when it is hot and evaporation is high. The roof "soil" does hold vast amounts of water.

Advantages of a green roof include:

- Reduction in heating and cooling costs. On hot days, the surface of a green roof can be cooler than open air, whereas the surface of a conventional roof can be up to 50°C (90°F) warmer.

- Retention of stormwater runoff. The soil and plants absorb significant rainwater, instead of immediately directing it into stormwater drains.

- Uptake of carbon dioxide and pollutants from the atmosphere, and of heavy metals, acid precipitation, and other rainwater pollutants out of rainwater.

- Honeybees and other insects thrive on the plants, attracting birds.

- Roof plants are beautiful and change color with the seasons.

- Describe the current condition of vegetation on the green roof.

- What could go wrong if the green roof is not properly maintained and periodically weeded?

More Brooks Hall green roof information can be found at this state web site:

WV Department of Environmental Protection, West Virginia University Green Roof, https://dep.wv.gov/WWE/Programs/stormwater/MS4/green/WVU/Pages/default.aspx (retrieved May 2019).

The Brooks Hall green roof is the last stop on the trip. **Do not forget to retrieve any items you left in the lab.**

Biological Diversity

Reading Assignment that should be completed <u>BEFORE</u> Lab 10

Readings

Strahler, A.H., 2013, **Introducing Physical Geography**, John Wiley and Sons, New York, **Chapter 8,** *or alternative reading assigned in GEOG 107.*

Materials Used in Lab

• Hand Calculators (provided in lab), glossary in Strahler textbook.

Learning Objectives

Upon successful completion of this lab, students will be able to:

• Take part in a study, including data tabulation, analysis, and discussion
• Calculate species richness
• Define the species-area relationship
• Visually identify spatial patterns of clustering

Species Richness

Biodiversity, once a technical term used only by ecological scientists, is now part of every-day language. In scientific use, Biodiversity means the number of species in a study area, either total species or species within a specified group, such as "plants", "mammals", "insects", *etc.* There are many different ways to measure the number of species, but the most common method is **species richness**, the number of different species existing in the study area.

Species-Area Relationship

One biogeography theory, the species-area relationship, holds that the number of different species increases as larger and larger areas are sampled. There are many possible reasons behind this biodiversity phenomenon.

1. Briefly give three explanations for the species-area relationship.

White Park Study Area

White Park is a city park with a history of human impact. First cleared for farms, the area was an oil storage site in the late 1800s to early 1900s (Figure 10-1), when earthworks were built to contain oil spills and towering fires were caused by lightning. It has reforested since the mid-1900s, but relict earthworks make an unusual topography with scattered high, dry locations and low, wet sites. The forest is dominated by sugar maple, bitternut hickory, black cherry, and yellow poplar. Most trees are 60–70 years old; but a few 100+ year old trees remain.

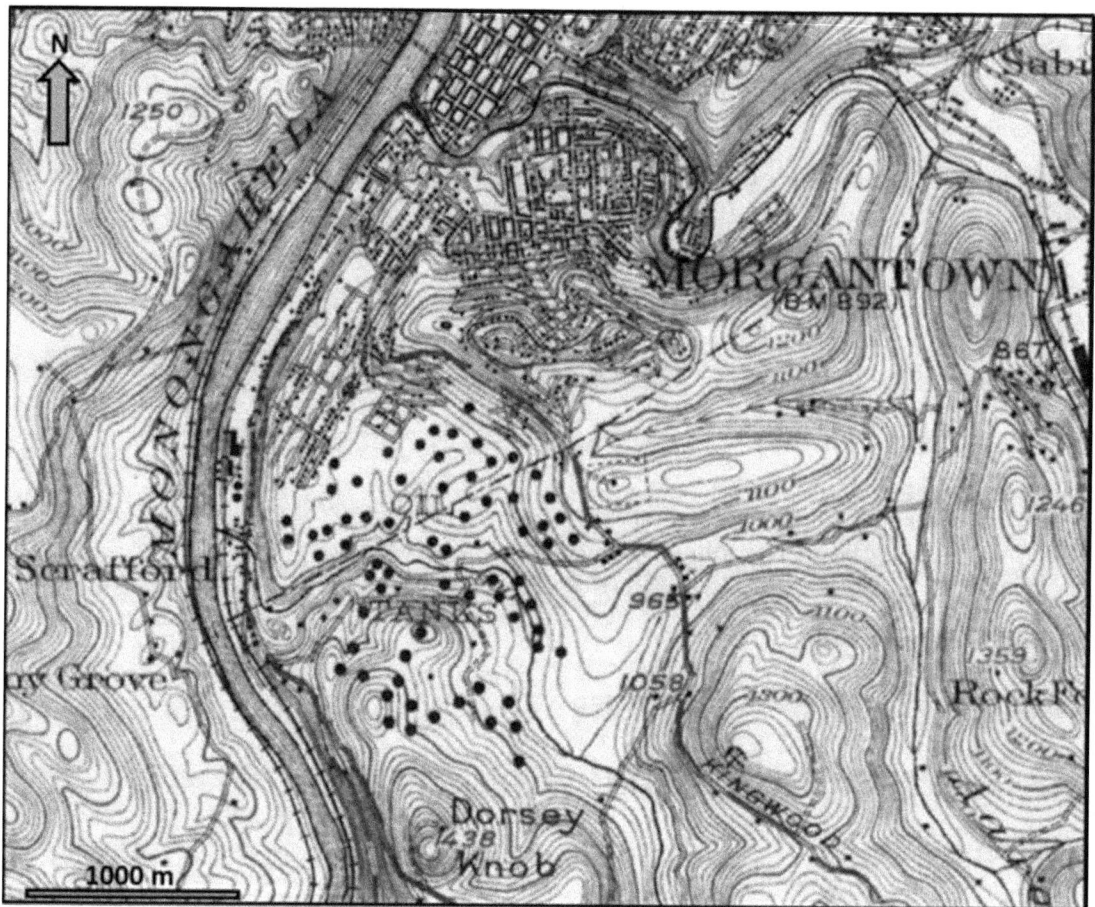

Figure 10-1   *A 1925 Morgantown map (retrieved from USGS topoView site at https://ngmdb. usgs.gov/topoview/, June 2018), shows circular oil storage tanks.*

Nested Quadrat Sampling Design

Biogeographers use nested **quadrats** (habitat sample areas) to investigate species-area relationships for study locations. Quadrat shape varies, but always represent a number of area units "nested" within one another (Figure 10-2).

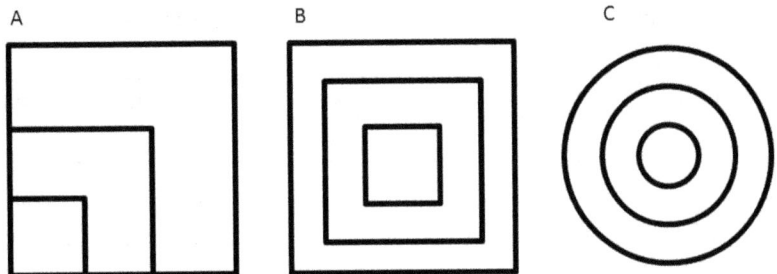

Figure 10-2 *Nested sampling designs made up of square quadrats (A & B) and circular quadrats (C).*

Study Design

In this exercise, we will use data to examine how species richness varies with quadrat size in a human-impacted area. The data were collected in White Park.

The study used a nested quadrat design, with the following quadrats dimensions:

- Quadrat 1 = 2 m × 2 m

- Quadrat 2 = 5 m × 5 m

- Quadrat 3 = 10 m × 10 m

- Quadrat 4 = 20 m × 20 m

- Quadrat 5 = 30 m × 30 m

Data Analysis

Quadrat 1 is the smallest quadrat and is contained in all other quadrats. Quadrat 2 is the second smallest quadrat and is contained within quadrats 3, 4, and 5.

2. Use the table below to list plant species in each quadrat based on map data in Figure 10-3. Remember all species in quadrat 1 also occur in subsequent quadrats, all species in quadrat 2 also occur in subsequent quadrats, *etc.*

Quadrat 1	Quadrat 2	Quadrat 3	Quadrat 4	Quadrat 5

3. Fill out the species richness and calculated area of each quadrat in this table.

Quadrat	Species Richness	Quadrat Area (m²)
1		
2		
3		
4		
5		

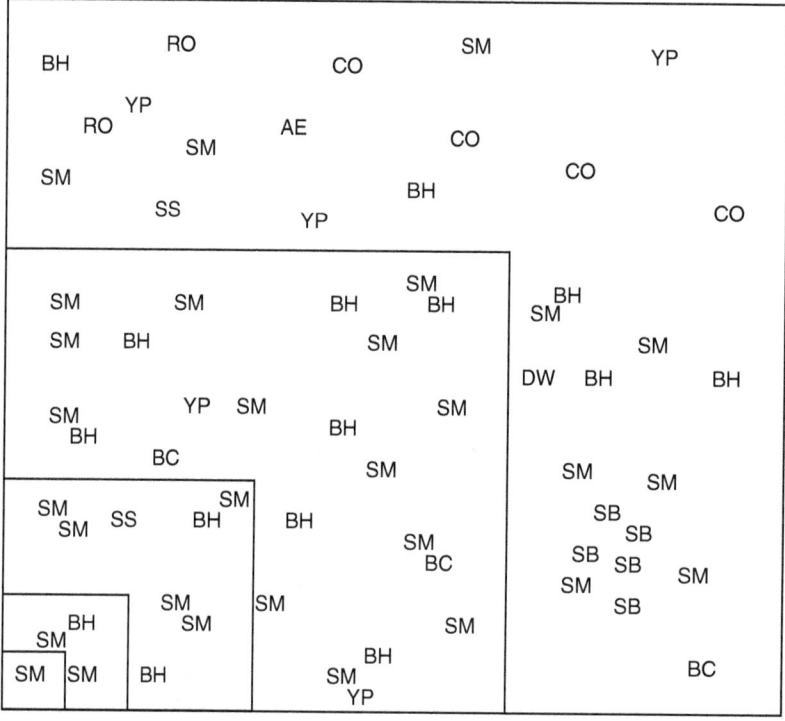

Tree Species Key

SM – sugar maple   BH – bitternut hickory
SS – sassafras   BC – black cherry
YP – yellow poplar   SB – sweet birch
CO – chestnut oak   RO – red oak
DW – dogwood   AE – American elm

Figure 10-3 *Map view of the study site, the nested quadrat pattern, and tree species present represented by initials given in the key below.*

4. Plot a species-area curve for the White Park tree data in Figure 10-4 below. Plot quadrat area (the independent variable) on the horizontal axis and species richness (the dependent variable) on the vertical axis. Label each axis and make tick marks to denote area and number of species. Consider plotting area as a logarithmic function (1, 10, 100, 1000 m²).

Figure 10-4 *Graph showing how species richness varies with sample quadrat area in White Park, Morgantown, West Virginia.*

Results and Discussion

5. Does the species-area relationship hold true for White Park?

6. Describe any spatial patterns that appear on the original map (Figure 10-3). For example, are certain species clustered together or spread apart? If so, which ones tend to cluster together?

7. Why might clustered spatial patterns occur?

8. How might these patterns and their possible causes impact the apparent relationship between species diversity data and sample quadrat area?

9. In what types of sites would you expect a strong species-area relationship?

10. In what types of sites would you expect a weak species-area relationship?

Lab Exercise 11

Watershed Research Using GIS

Assignments that must be completed <u>BEFORE</u> Lab 11

Readings

Strahler, A.H., 2013, **Introducing Physical Geography**, John Wiley and Sons, New York, **Chapters 1 (p. 21–33) and 14**, *or alternative readings used in GEOG 107*.
Richard, Glenn A., 2018, What is Google Earth? https://serc.carleton.edu/sp/library/google_earth/what.html (retrieved May 2019).

Materials Used in Lab

Access to Google Earth
Digital **GIS files available only on GEOG 106 lab computers** in Brooks Hall

Learning Objectives

Upon successful completion of this lab, students will be able to:

- Distinguish between points, lines, and polygons in a GIS
- Create points, lines and polygons in Google Earth
- Distinguish between raster and vector GIS data
- Insert an image into Google Earth
- Estimate error in a GIS
- Identify a variety of land cover types using a digital orthophoto

Before starting the lab:

- **You must complete Lab Exercise 3:** Google Earth for Physical Geographers **before attempting this lab**.
- **Review Lab Exercise 3:** Google Earth for Physical Geographers.
- **Make sure you understand** these GIS terms: **GPS, GIS, spatial object, point, line, polygon, wetland,** and **watershed**.

Look up unfamiliar terms in the USGS <u>Water Science Glossary of Terms</u> at http://water.usgs.gov/edu/dictionary.html (retrieved June 2018) or your GEOG 107 textbook glossary.

Google Earth and GIS Fundamentals

Log in to a Brooks Hall teaching lab computer, and start Google Earth by locating and double-clicking the Google Earth Icon on the desktop or selecting Google Earth from the list of programs in the 'Start' menu.

If you are not very comfortable using Google Earth, go back over what you did in Lab 3. Detailed instructions like those given in the Lab 3 are helpful for starters, but eventually the best way to master Google Earth is by playing with the software.

Locate your home using navigation tools you learned in lab 3 (e.g. the mouse 'grab' pan tool, the mouse scroll button/wheel, the navigation tools on the right side of the display, or entering a location in the 'Fly to' tool). Try several tools for practice.

Point Features

A **point** is one of three types of **vector data** used in Geographic Information Systems (GIS). The other two types of vector data are **lines** and **polygons**. Later, we will use non-vector **raster data**, but we will focus on points, lines, and polygons for now.

The pushpin symbol for a placemark creates a point location wherever you direct.

Using the Placemark icon in the banner at the top of the screen, create a yellow pushpin Placemark *point* precisely at your home.

A 'New Placemark' box pops up when the 'Add Placemark' icon is selected. When the box is active, first give the placemark a name and then move it, change its color or style, and add a description as appropriate. To place a placemark in its correct location, you may have to move the 'New Placemark' box aside, taking care to keep it open.

A closed placemark box can be reopened in an 'Edit Placemark' box by right-clicking either the pushpin or the placemark name listed under 'Places' in the left-hand frame, then drawing down to 'Properties'.

1. Reading from the placemark box, enter your home's latitude and longitude in decimal degrees. (Remember you may have to use *Tools > Options > 3D View > Decimal Degrees* to reset the coordinate format.)

   Latitude _____ Longitude _____

If not already done, open the 'Edit Placemark' box and fill in a description. Explore the other tabs, and make a few changes to see how these features work.

2. What is your favorite style and color for a placemark pushpin symbol?

Close the placemark box when you finish changes by clicking OK or the close X.

The placemark will still appear in the 'Places' the left-hand frame box. When you double-click on this or any other placemarks, Google Earth will 'fly to' that location.

3. Zoom out, relocate to a new city, and fly back to your placemark. Isn't this fun?

If you want to delete the placemark, select it, right-click with the mouse, and draw down to 'delete'. You can even email placemarks. (*Try it later if you have time!*)

Line Features

The term 'line' is used more broadly in GIS than the narrow definition you may remember from a math class. We consider any unclosed feature consisting of line segments as a 'line' feature.

Many vector features, both natural (i.e. streams) and artificial (roads), can be symbolized by lines. A line feature is called a 'path' in Google Earth. You may also have seen line features referred to as 'routes' or 'tracks' if you navigate using GPS, a Global Positioning System.

4. Create a path for your walk to class, a favorite running route, the trip between home and a popular dining establishment, or some other familiar path meaningful to you.

To create a path in Google Earth, click on and select the 'Add Path' tool, the fourth button from the left on in the toolbar button menu.

Up pops a 'New Path' tool box, which works like the 'New Placemark' box. Keeping it open, name the path, then move the mouse over the screen and see the cursor become a crosshair symbol. You can left-click to create individual line nodes one at a time or use a streaming mode by clicking and holding the left mouse button. When the path is competed, move the cursor to the box, add a description, edit symbols, and click OK.

That's it. You've created a path, a 'line' feature. Now close it.

Measuring Line Features

Either click on the line feature you just created or click on its path name in the left-hand frame and drawing down to 'Properties'. Either step will open an 'Edit Path' box. Click on the Measurements tab to reveal the length of the line you created.

5. How might the precision and accuracy of the clicks you used to create the path influence its measurement?

6. How might the length of this path vary if you followed virtually the same path but used only half as many clicks to create it?

7. How might path length vary if you used ten times more clicks to create it?

The 'Measure' tool, show by a ruler in the icon menu, is another way to measure linear distances between points, following either straight lines or multiple-node paths. If you are not comfortable with the tool from lab 3, we encourage you to try it out again.

Polygon Features

Polygons represent areas, and can take any shape. Creating a polygon is similar to creating a path, and is accomplished using the Polygon tool, the six-sided symbol to the immediate left of the Path tool.

Use the Google Earth search tool to fly to Mountaineer Field at Milan Puskar Stadium, a good place to learn how to create a *polygon* feature. Click on the Polygon tool to open a 'New Polygon' box, set polygon units to 'yards', and create a rectangular polygon outlining the football playing field: just the area between the sidelines and end lines. Don't include seating sections or any part of the field that is out of bounds.

Like in creating a path, one can left-click on individual points one at a time or use the left mouse in streaming mode. Single point clicking is more precise, but slower. Things can really get out of control using the faster streaming mode. Individual points in a polygon can be moved, added, or deleted when a 'New Polygon' or an 'Edit Polygon' box is open. Editing polygons made in streaming mode can take a lot of effort.

8. Click on the Measurements tab to see the calculations, and enter them here:

   Perimeter _____ yards   Area _____ square yards

9. O.K. football fans! What are the real dimensions, based on NCAA field regulations?

   Perimeter _____ yards   Area _____ square yards

10. How close were your measurements to the dimensions of a regulation field? List possible sources of path measurement error. How could you do better next season?

High-Elevation Appalachian Watersheds

We can use Google Earth to explore some of the highest and wildest areas ecosystems in the Central Appalachians, areas among the very last places in the eastern United States to be settled by European Americans. The highest valley of its size east of the Rocky Mountains, Canaan Valley is a National Natural Landmark because of its unique high-elevation plant community; most of the valley is within the Canaan Valley National Wildlife Refuge. The Upper Blackwater River hydrologic unit covers most of Canaan Valley, while the Lower Blackwater River hydrologic unit to the west contains scenic Blackwater Falls and the rugged Blackwater Canyon. Together, the two hydrologic units make up the Blackwater River Watershed. Nearby Red Creek watershed, southeast of the Upper Blackwater unit, drains much of the Dolly Sods Wilderness Area. All three of these remote, biologically diverse areas are very popular with day hikers, backpackers, and biogeographers.

USGS Hydrologic Unit Codes (HUC)

The USGS has devised numerical designations for the national water data network. In West Virginia, the first two digits of the designation are either 05 for streams that ultimately drain into the Ohio River, or 02 for streams that feed Mid-Atlantic drainage. Four-digit subdivisions include the 0505 for those that drain into the New-Kanawha River system *vs.* 0502 for streams that feed the Monongahela River.

The Cheat River sub-basin is designated with 8 digits 05020004, which includes the 10-digit 0502000402 Blackwater River watershed. Up to 12 digits are used as watersheds get smaller and smaller. West Virginia's 12-digit watersheds range from 10,000 to 40,000 acres (4,000 to 16,000 hectares).

Going beyond what we need to know today, if you want to go into hydrology or geographic information science, the following web resources may be use useful.

Youngman, B., and Dahlman, L., 2018, What's a Watershed? Part B: Explore Your Watershed in Google Earth, https://serc.carleton.edu/eslabs/drought/2b.html (retrieved June 2018).

West Virginia GIS Technical Center, 2018, Watershed Boundary Dataset (8, 10, and 12 Digits), http://www.wvgis.wvu.edu/data/dataset.php?ID=123 (retrieved June 2018).

Exploring the Appalachian Watershed Using GIS

Now that we have reviewed Google Earth and become acquainted with the Blackwater River Watershed and HUCs, we will begin to see some of the power that the GIS software has to offer by studying the Blackwater River Watershed.

Google Earth and some other GIS software systems use layers stored in large files formatted in Keyhole Markup Language—**KML files**. The vast amount of data in many KML files lead to them usually being stored as compressed Keyhole Markup language, Zipped—**KMZ files**.

Proceed by loading the West_Virginia_12_digit_Watersheds.kmz file into Google Earth.

- Click the File tab in the main Google Earth menu, and then click the Open tab.

- Browse to and click the West_Virginia_12_digit_Watersheds.kmz file.

- Click the Open button (or double-click on the file; either method opens the file).

The large file takes time to load, but in a few seconds, the 12 digit watersheds appear on screen. Note the watersheds listed in the Places menu.

Using the West Virginia 12-digit watershed list in Places, scroll down to find the Upper Blackwater River. Double-click it, and zoom in to the polygon outlining the Upper Blackwater River watershed.

11. Using the measure tool button to find the length and width of the watershed.

   Approximate length: _____ km.   Approximate width: _____ km.

Now that we have examined the drainage in two dimensions, we will next view the watershed in three dimensions, by adding elevation or 'relief'. In 3D, you can see the drainage basin boundaries that define the Upper Blackwater River watershed.

Turn on 'Use 3D Imagery' by clicking through 'Tools' and 'Options', and set Elevation Exaggeration = 3. Click 'Apply', and then close the Options box by clicking 'OK'.

If you are not currently viewing the entire Upper Blackwater River watershed on your screen, do so using the zoom tool.

If your mouse has either a middle button or a depressible scroll wheel, tilt the view by depressing the button and moving the mouse forward or backward. If your mouse has a scroll wheel, you can tilt the view by pressing the Shift key and scrolling. You can also press Shift, the left mouse button, and drag.

Return to the 'nadir' or directly overhead viewing angle, using reset or the mouse and zoom tool. The entire Upper Blackwater River watershed should now fill your screen.

Raster Data

In contrast to the vector data of points, lines, and polygons, **Raster Data** are picture element (pixel) based. Our billions and billions of digital photographs stored on our smart phones and posted on social media are all comprised of raster data—each image made of hundreds of thousands to tens of millions of information bytes!

Raster data can make very powerful additions to a GIS by allowing one to incorporate layers such as satellite imagery, aerial photographs, weather data, geological maps, soils maps, historic maps, land-use maps, etc. In fact, the Google Earth satellite base layers are composed of raster data.

Our next step is to add a raster image of the Blackwater Falls 1:24,000 scale 7.5 minute topographic quadrangle to our GIS. We will provide the image in lab, but several different types of raster images of a 2016 version of this quadrangle and virtually every edition of USGS quadrangle ever published can be downloaded from the USGS topoView National Geologic Map Database (https://ngmdb.usgs.gov/topoview/viewer/).

There are two ways to add raster imagery to Google Earth. One is via manual adjustment of the image boundaries using natural and man-made features as references to line up (= 'co-register') the two images. This can be difficult and time consuming depending on the image, but is necessary when the overlay image does not have known coordinates.

The simpler alternative method is to enter the known coordinates for the four corners of the image, and let Google Earth overlay the two images. We will use this method.

12. Change latitude & longitude units to display in **"degrees, minutes, seconds"**.
   a. Use the main Google Earth menu to browse to 'Tools>Options>3D View' and make sure the 'Show Lat/Long' selection is 'Degrees, Minutes, Seconds'.
   b. Close the box.

13. Position the Google Earth image
   a. Position the image so the "Courtland" place name is near the middle of the screen.
   b. On the main Google Earth screen, zoom in or out so the scale bar at the bottom left of the screen reads between 5 and 6.5 km (3 and 4 mi).

14. Click the 'Add' tab on the top menu bar, and select 'Image Overlay'.

A rectangular image placeholder, outlined in green, will appear as will a 'New Image Overlay' box that has a browse button with a link, a transparency slider, and five tabs.
   a. Click the 'Browse' button, and browse to and **select** the **blackwater_falls.gif** file.
   b. Enter "Blackwater Falls 7.5 Minute Quadrangle" in the 'Name' line,

15. Click the 'Location' tab, and carefully enter the following coordinates:

| North: 39° 7'30.00"N | East: 79° 22'30.00 **W** |
| South: 39° 0'00.00"N | West: 79° 30'00.00'"**W** |

Click 'OK', and the Blackwater Fall 7.5 Minute USGS Topographic Map should show up over the underlying Google Earth base image.

16. Adjust the transparency bar back and forth to see how the overlay aligns with the base Google Earth image.

The newly added topographic map image name will now appear in the "Places" menu on the left side of the screen. If you select it, right-click, and draw down to the 'Properties' tab, you can edit the boxes and tab settings in an 'Edit Image Overlay' box.

Horizontal Error

Error is inherent in any Geographic Information System. You may have experienced it yourself if you have used a GPS and observed the position of your car has wandered off the mapped road! It is critical to estimate this error and be able to report it.

To estimate **error** in your georectified this image, use the slider tool below the image overlay to see how well it lines up with the underlying Google Earth base image.

17. Compare the following features and describe how well the topographic map is georectified (geographically positioned). Estimate the error in your georectified image by measuring the offset between fixed locations identifiable on both images.

Common point on both maps	Distance error (km)
Crossroad of County Rt. 32/14 (Canaan Heights Rd.) and County Rt. 32/15 (Beardon Knob Rd.)	
Intersection of WV Rt. 32 (Appalachian Hwy.) and County Rt. 32/16 (Timberline Rd.)	
Intersection of WV Rt. 32 (Appalachian Hwy.) and WV Rt. 72 (Dry Fork Rd.)	
Confluence of North Branch with Blackwater River, east of Canaan Heights	
Average of 4 error measurements	

Delineate Different Land Cover Types in the Blackwater Watershed

18. You will notice that different types of land cover have different colors and textures on the image. Identify these cover types and describe the color and texture of the following features:

Golf course: _____

Spruce forest: _____

Hardwood forest: _____

Housing development: _____

Ski area: _____

Wetlands: _____

Using the polygon tool, create polygons for all obvious wetlands. Save each one with a different name so you can add up these measurements later.

19. How many wetlands occur in the Upper Blackwater Watershed?

Canaan Valley and the Upper Blackwater River watershed has the largest wetland area of any location in the Central Appalachians.

20. Adding the areas shown on the 'Edit Polygon' measurements tab, calculate the approximate total area covered by wetlands in the map area: _____ km^2.

9 781524 996369